MARITAL
CHOICES

Also by William J. Lederer

The Mirages of Marriage (with Don D. Jackson)
The Ugly American (with Eugene Burdick)
A Nation of Sheep
All the Ship's at Sea
Ensign O'Toole and Me
Timothy's Song
Sarkhan (with Eugene Burdick)
Spare-Time Article Writing for Money
The Story of Pink Jade
The Last Cruise
Complete Cross-Country Skiing and Ski Touring
Marriage for and Against (contributer)

MARITAL CHOICES

Forecasting, Assessing,
and Improving a Relationship

BY *William J. Lederer*

W·W·NORTON & COMPANY
New York   *London*

Published simultaneously in Canada by George J. McLeod Limited,
Toronto. Printed in the United States of America   All Rights
Reserved   First Edition

Grateful acknowledgment is made to the following: The
Bobbs-Merrill Company, Inc., for permission to reprint an excerpt
on the Coca Pulse Test from *The Food Connection*, Copyright 1979
by Dr. David Sheinkin, Dr. Michael Schachter, and Richard Hutton;
and Larchmont Books for permission to reprint excerpts from Dr.
Harvey M. Ross' *Fighting Depression*.

Library of Congress Cataloging in Publication Data
Lederer, William J   1912–
Marital choices.
1. Marriage.   2. Interpersonal relations—Problems, exercises, etc.
I. Title.
HQ734.L3753   1981   306.8′7   80–21259
ISBN 0–393–01412–6

W. W. Norton & Company, Inc. 500 Fifth Avenue, New York, N.Y.
10110
W. W. Norton & Company Ltd. 25 New Street Square, London
EC4A 3NT

1 2 3 4 5 6 7 8 9 0

Contents

PART II
THE PHYSIOLOGICAL FACTORS   201

William J. Lederer is a member of:
The International Academy of Preventive Medicine
The Academy of Orthomolecular Psychiatry
The International College of Applied Nutrition
The European Academy of Preventive Medicine

MARITAL
CHOICES

About This Book

There are good marriages and there are bad marriages. The spectrum goes from heavenly accord down to hellish discord. However, be the relationship good or bad, all of them have one common quality.

Those partners who have a bad marriage have chosen—have actually contracted—to have a bad marriage. And even though they may protest their condition, they still make daily choices to maintain the miseries of their relationship.

Likewise a good marriage. The partners who have a good marriage have chosen—have actually contracted—to have a good marriage; and they make daily choices to build their relationship into an even better one.

The activist-philosopher Gurdjieff taught that like attracts like, and that an individual will go through life attracting the same kind of scenes, problems, and people until that individual changes (this means, of course, until the individual changes his or her behavior).

These, then, are the marital choices: By choosing to change your behavior you can choose to change your marriage; however, since there sometimes are physiological problems which make a change in behavior difficult, here again the choice can be made for improvement. These are the choices with which this book is concerned. The book is for those couples who want to make upward choices, who are sufficiently eager to improve their relationship that they are willing to modify their behavior and, if necessary, work to improve their physical health in order to do so.

What follows is neither an encyclopedia nor a general textbook on marriage. It is a facilitator, a training manual for wives and husbands who want to transform a bad marriage into a satisfying one or make an acceptable marriage even better.

It is also a guidebook for unwed couples who want to improve their relationship or assess their chances of having a gratifying relationship if they marry.

In part, this is a do-it-yourself manual designed for couples who want to improve their relationship by themselves, privately. However, it is also for those who prefer working with professional therapists or who would like their marital improvement program monitored by a friend, a member of the clergy, or a para-professional.

The texts and exercises are designed so that any of these approaches can be accommodated.

This manual is not an argument for or against marriage as an institution. It has only one mission: to assist couples, *to the degree that they want to be assisted,* in reducing their levels of discontent and elevating their levels of relationship gratification.

We cannot promise a relationship devoid of problems. No therapy, no program can elevate a marriage to a state of constant peace. Such a condition does not exist in life, and if it did, it probably would expire from boredom.

However, most spouses who make the effort this book requires can develop the capabilities needed to build a good marriage. They will learn to visualize the nature of a good marriage. They will be exposed to the processes common to all satisfying relationships. Then, *if they both desire,* the book can lead them, step by step, into acquiring the skills which can produce these processes (a good relationship).

Although not every endangered couple reading this book will escape divorce, most spouses, even those who have considered divorce, can benefit from its program.

Similarly, we cannot assure unmarried people that, as a result of this book, they will choose the perfect mate and live happily ever after. However, we are confident that unwed couples who go through the texts and exercises together can gain a perspective which will help them see through the chimera of raw romance. They will be more able to assess the viability of a long-range and satisfying relationship. And, if they decide to marry, they will start with some essential tools: the behavioral skills, the attitudes, and the negotiation facility which will help them cope with the realities of married life.

Fundamentals

We hold that every marriage is unique because every person is unique. Therefore, we offer no rigid blueprint for *the* perfect marriage. Instead, we assist readers in defining their particular needs and satisfactions—the ones which, when realized, will create a satisfying marriage.

OUR APPROACH

There are no directives in this book. We consider ourselves the colleagues and assistants of this book's clients. The instructions are presented so that couples can develop their own interpretations at their own pace and thus improve their relationship in small, easy steps through study and training exercises at home.

WILL A NEW HAPPY MARRIAGE LAST FOREVER?

We believe that two individuals and their unique long-range relationship, like everything else in life, constantly change. The behavioral interactions which produce joys today may become obsolete with the passage of time or the development of new circumstances. Therefore, we show readers how to become aware of the changes within themselves and their lives together, and how to adjust so that the levels of satisfaction remain high.

THE ESSENCE OF THE PROGRAM'S CONTENT

The major part of the book, Part I, consists of behavioral and psychological exercises for spouses to do together at home. Even if a professional therapist is involved, we prefer the spouses to do the exercises in the intimate environment of their home, and, frequently, with their children present and participating. Part II consists of a discussion of physical conditions which can influence marriage, and of how to identify and cope with these conditions.

THE REVOLUTIONARY NATURE OF THE PROGRAM

Some of the training exercises in this manual are revolutionary. They were designed to bring about radical change, to change marital dissonance into marital accord. The mission of the exercises is to fragment negative behavioral patterns which have been rehearsed for years and, simultaneously, to assist spouses in replacing the old patterns with positive, satisfying ones.

Achieving this is within the capabilities of most spouses *when both earnestly* desire to achieve it.* We will present case histories of spouses negotiating pacts which brought peace to the relationship. They start with each frankly stating what each needs and wants from the marriage. Some of the initial requests may insult various gender biases. Some may provoke angry thoughts like, "Oh, that's the absurd behavior of a male chauvinist pig!" or "Oh, that's the absurd behavior of a female chauvinist sow!"

It must be remembered that sexism does exist and that partners in the case histories were motivated by their unique needs and desires. It may be that some of *your* needs and desires would appear absurd indeed to other couples. Let everyone, however, mind his or her own business.

It is the *process* of working out these idiosyncratic needs and desires (so that both spouses have maximum satisfaction) which is the good revolution. It is the *process* which is important, not the content. The content is valuable only as an illustration of the process at work. Organizing your marital processes (your behavioral interactions) is like creating your own Bill of Rights, your own Constitution. With this type of revolution it is possible for both parties to be satisfied. Everyone is a winner. In marriage there never is one winner and one loser. There are either two winners or two losers.

---

*The word *earnestly* has a beautiful symbolism here. It means "willing to pay a price now for something which will be received in great abundance later."

This book is designed to help you to have a marital revolution which will produce two winners.

FACTORS WHICH INFLUENCED US TO EMPHASIZE THE PROCESSES OF MARRIAGE (THE OBSERVABLE BEHAVIORS), AND GENERALLY TO DE-EMPHASIZE THE CONTENT OF THE MARRIAGE

At the Behavior Research Institute in Peacham, Vermont, before that at the Mental Research Institute in Palo Alto, California, and at various times with colleagues such as Dr. Don D. Jackson, Dr. Norman Paul, and Dr. Richard Stuart; and in courses under Professors Percy Bridgeman of Harvard* and Norbert Wiener of MIT†, we objectively have observed and analyzed the marital praxis. We have identified factors which appear common to most good marriages and also factors which appear common to most bad marriages.

It is from these observations, much study, and considerable clinical experience that in this book we propose broad, flexible actions for marital improvement. These actions are applicable to most couples regardless of their individual and mutual idiosyncracies. They also are applicable to most families regardless of their cultural and ethnic background.

Some couples who want to improve their relationship do not respond to even the most competent psychological therapy. There can be many causes for the blockage. However, clinical evidence suggests that a significant number of these unhappy couples suffer from subtle physical pathologies of which they may not be aware. It is possible that physical malfunctions and nagging discomforts can provoke negative behaviors.

Because this problem does not affect all readers, we will supply the appropriate information in Part II. However, for those of you (either one or both) who appear to be bothered by chronic fatigue, periods of unexplainable irritability, or unpre-

---

*The Application of Mathematics to Social (Family) Problems.
†Cybernetics in Marriage.

dictable irrational behaviors, we suggest that you turn to Part II (page 215) now and take the test, to see whether you need the information supplied there on some of the newer findings and theory in this field.

Preliminaries

ABOUT OTHER MEMBERS OF THE FAMILY

Improving a marriage involves all the concerned parties. There may be children. They are concerned parties and therefore must be considered. Throughout the program we will indicate which exercises are appropriate for the children's participation.

If the spouses start to change their relationship without the participation of the children, there may be trouble. Concerned about their role in the changed relationship, children often take action to maintain the status quo. One of their methods is to play one parent against another. They struggle often to retain a familiar and predictable environment. They may even seek to resist the improvement of a marital situation.

The family is a system, and all systems tend to resist change unless all parts of the system participate in the change. In Part II there is a detailed discussion of the systems theory of human relationships.

CONCERNING THE BEHAVIOR OF QUARRELING SPOUSES WHEN THEY START THE PROGRAM

Both partners should refrain from hinting about or mentioning separation or divorce during the five weeks of the program —starting now. Also, we urge spouses to avoid all "I'll get even" actions while working with this manual.

We are not suggesting that you stop *thinking* about negative actions or about separation or divorce. It is almost impossible to control one's thoughts. However, you can stop yourself from talking or acting on your negative fantasies or threats.

Disciplining yourself in this manner may require considerable effort. We urge you to make the effort. Five weeks of self-discipline is a wise investment to overcome the effects of earlier years of marital quarrels or a lifetime of negative conditioning. You now have the opportunity of escaping the chains to the past and of building constructively for now and for the future.

SCHEDULING THE PROGRAM

The marital improvement program is designed for busy people. It can be carried out by spouses having full-time jobs and a house filled with children.

On the average, each training assignment requires about one hour per day. The sum of them takes about five weeks. The exercises can be done at any agreed-upon time, which can vary from day to day.

If you want to invest more than an hour a day on an assignment, that's fine. If you need to spend more than one day on an assignment, that's also fine. *However, do not complete more than one assignment a day.* Much research and experience indicate that carrying out a maximum of one assignment per day results in optimal progress.

KEEPING RECORDS

Some of the training exercises require simple paperwork. Save all of your paperwork—preferably in one notebook kept by both spouses. You will find it helpful—and often necessary—to refer to your notes as you move ahead in the program.

THE SCOPE OF THE PROGRAM

The principles and most of the exercises, in both Section I and Section II, can be applied successfully to any relationship involving communications and negotiations, including business relationships.

Part I

A FIVE-WEEK
PROGRAM
TO IMPROVE
ANY TYPE
OF RELATIONSHIP

Assignment 1

In this first assignment, you will take an inventory of your marriage—mostly to establish how each partner assesses and perceives the relationship. The function of the marital inventory is to identify the areas in which changes should be made. It is not designed to prove who is right and who is wrong.

Also, you will begin a short ritual to initiate intimacy and friendly communication. This is scheduled for every day of the program.

THE ASSIGNMENTS

1. Take a marital inventory. (The test is to be taken separately by each partner.)

2. Start the Intimate Time ritual.

Taking a Marital Inventory

The test is to be taken separately by each partner. A glance at the actual test, which starts on page 27, will show that it consists of six different clusters, each of which concerns a specific element of your marriage.

CLUSTER 1 concerns your commitment to your marriage

CLUSTER 2 concerns the effectiveness of your communication with your spouse.

CLUSTER 3 concerns the "cherishing behaviors" which set the tone of the marriage.

CLUSTER 4 concerns your performance in the everyday at-home behaviors.

CLUSTER 5 concerns the way you make decisions within the marriage—thus determining the power structure of the relationship.

CLUSTER 6 concerns the way you respond to marital conflicts.

Note that each cluster has seven questions. Both spouses answer the same questions on several different subjects within each cluster.

Each item is answered independently of the others. The answer is indicated by circling one of the numbers. Base your answer on your general assessment of your feelings over the past seven to ten days. In other words, do not consider the effect of any recent, unusually positive or negative experiences.

For any item to which you select a low answer of 1 or 2 (giving your spouse a low grade), you must *write* the following two things:

1. A positive and specific request for a change in that particular behavior of your spouse.

2. What you are willing to do to assist your spouse in making the changes you have requested of her or him.

What you have written is essential for this assignment. It also will be useful to you in later exercises. After the marital inventory has been completed, put this information—along with all other paperwork—in your common notebook.

Remember, each spouse is to take the test separately.

When all questions have been answered, couples will score and interpret their responses together.

AN EXAMPLE OF HOW TO DO THE TEST

In the first cluster of questions, on Item 1(a), John has rated Mary with a low grade of 2 as follows:

1(a)   *I enjoy spending time with my spouse.*

   1   ②   3   4
Rarely true   Almost always

The reason John gave Mary a low rating of 2 was that in John's opinion Mary was an incessant talker, never letting anyone else talk. It seemed to him that she talked interminably, repeating herself over and over again. This bored John and he seldom listened to her. Mary, as a result, complained that he paid no attention to her and thus was disrespectful. They frequently quarreled over this situation.

John read the instructions for this exercise and saw that he had to write a positive and specific request for Mary to change her talking habits (because he had given her the low rating of 2). He wrote:

"I request that Mary pause for a moment before she talks; and that during the pause she ask herself, " 'Is this necessary? Have I already said it before?' "

John also had to write what he could do to facilitate the change he desired in Mary. He wrote:

"I can assist Mary by looking directly at her and listening while she's talking. As a result of this, she eventually will know that I hear her and am interested in what she has to say. Perhaps then she will have no need to repeat herself and will be more discriminating."

ANOTHER EXAMPLE

Mary, in doing her test, has rated John on Item 1(d) with a low grade of 2 as follows:

1(d)   I feel proud to introduce my spouse to my friends and associates.

    1   ②   3   4

Rarely true   Almost always

She gave him a low rating of 2 because John had put on a lot of weight and was a sloppy dresser. To her, John looked slovenly. This embarrassed Mary. It also annoyed her because he had done nothing about it.

Having understood the instructions, Mary wrote a positive and specific request concerning John's weight. She wrote:

"I request that John see a doctor or join an organization like Weight Watchers so that he can start a physical fitness program and have a good figure again."

Mary also wrote what she could do to facilitate the change she was requesting John to make. She wrote:

"I can help John by having meals which are prescribed for him by the doctor or Weight Watchers. I'll eat them also. Also, I'd be willing to contribute a quarter of my personal allowance for several months to help him buy new clothes when he's back to normal weight."

Now make your own marital inventory.

A Marital Inventory

CLUSTER NO. 1
This series of questions concerns your *commitment* to your marriage.

    a. I enjoy spending time with my spouse.

        1   2   3   4

rarely true   almost always true

    b. I am willing to set my own pleasure aside in order to do something which is important for my spouse.

        1   2   3   4

rarely true   almost always true

    c. I have positive thoughts about the prospect of spending the coming years with my spouse.

        1   2   3   4

rarely true   almost always true

    d. I feel proud to introduce my spouse to my friends and associates.

        1   2   3   4

rarely true   almost always true

    e. I feel that I would marry the same person again if I had the opportunity to do so.

        1   2   3   4

rarely true   almost always true

    f. On balance, I am considerably more happy than unhappy in my marriage.

        1   2   3   4

rarely true   almost always true

    g. I feel that my marriage helps me to grow as a person.

        1   2   3   4

rarely true   almost always true

CLUSTER NO. 2

This series of questions concerns the effectiveness of your *communication* with your spouse.

    a. I feel satisfied when I discuss my worries and concerns with my spouse.

1	2	3	4
rarely true			almost always true

    b. My spouse and I ask for what we want from each other.

1	2	3	4
rarely true			almost always true

    c. My spouse derives considerable satisfaction from sitting down to talk things over.

1	2	3	4
rarely true			almost always true

    d. I feel that my spouse understands the important things that I try to communicate.

1	2	3	4
rarely true			almost always true

    e. My spouse and I generally have something to say to each other, or we have relaxed silences during which neither of us is uncomfortable.

1	2	3	4
rarely true			almost always true

    f. I can tell accurately the way my spouse feels even when the feelings are not put into words.

1	2	3	4
rarely true			almost always true

    g. I feel able to say to my spouse anything which is important to me.

1	2	3	4
rarely true			almost always true

CLUSTER NO. 3

The following questions are concerned with the cherishing behaviors which establish the tone of your marriage.

a. I feel that I do a great deal to show my spouse that I care about him or her.

| 1 | 2 | 3 | 4 |

rarely true   almost always true

b. I feel that my spouse does a great deal to show me that he or she cares.

| 1 | 2 | 3 | 4 |

rarely true   almost always true

c. My spouse and I acknowledge the pleasant things which we do for each other.

| 1 | 2 | 3 | 4 |

rarely true   almost always true

d. When I ask for something that I want, my spouse openly accepts or rejects the requests, but I am still comfortable about asking.

| 1 | 2 | 3 | 4 |

rarely true   almost always true

e. I feel that the exchange of small, cherishing behaviors in our marriage is comfortably balanced.

| 1 | 2 | 3 | 4 |

rarely true   almost always true

f. I feel that the *number* of cherishing behaviors which we exchange is quite adequate.

| 1 | 2 | 3 | 4 |

rarely true   almost always true

g. I feel that the *range* of cherishing behaviors which we exchange reaches into most of the areas of my desires.

| 1 | 2 | 3 | 4 |

rarely true   almost always true

CLUSTER NO. 4

The questions in this section concern your performance in the everyday at home behaviors.

a. I feel that the division of labor in our marriage is equitable even though it may not be equal.

    1   2   3   4
rarely true   almost always true

b. I trust my spouse to follow through on his or her end of a bargain.

    1   2   3   4
rarely true   almost always true

c. I consistently keep my end of agreements made concerning chores at home.

    1   2   3   4
rarely true   almost always true

d. When we sit down to try to divide up the labor, I feel that my spouse and I both try to find the best mutual (two-winner) solution rather than trying merely to make great personal gains.

    1   2   3   4
rarely true   almost always true

e. The requests which my spouse makes are generally "fair" so that I do not feel that I am often confronted with situations in which I must say no.

    1   2   3   4
rarely true   almost always true

f. The agreements which I make with my spouse are sufficiently clear so that I know when I have done my share and when my spouse has done his or her part.

    1   2   3   4
rarely true   almost always true

g. Unless something extraordinary happens, I know from one week to the next what my workload will be so that I can plan how to use my free time.

	1	2	3	4	
rarely true					almost always true

CLUSTER NO. 5

The questions in this section concern the way in which you *make decisions* within the marriage. This process determines the power structure of the marriage—who has the authority.

    a.   I feel that I have authority to make independent decisions in the areas in which it is important for me to have autonomy.

1	2	3	4
rarely true			almost always true

    b.   I am comfortable about having my spouse make some decisions independently.

1	2	3	4
rarely true			almost always true

    c.   When we make a shared decision, we follow through as we have decided, without constantly opening up the matter for further discussion and backtracking.

1	2	3	4
rarely true			almost always true

    d.   I am well aware of my areas of responsibility and know, too, my spouse's area of responsibility.

1	2	3	4
rarely true			almost always true

    e.   I feel that when we have a shared decision to make, my spouse and I both give careful consideration to the other's desires.

1	2	3	4
rarely true			almost always true

    f.   I feel that when a decision is to be made, my spouse listens respectfully to my point of view.

1	2	3	4
rarely true			almost always true

g. My spouse and I usually feel that we have made the correct decision after most decisions are made: that is, we do not overlook important dimensions of problems which we attempt to solve together.

1	2	3	4

rarely true   almost always true

CLUSTER NO. 6

In this section, we ask about the way in which you respond to conflict.

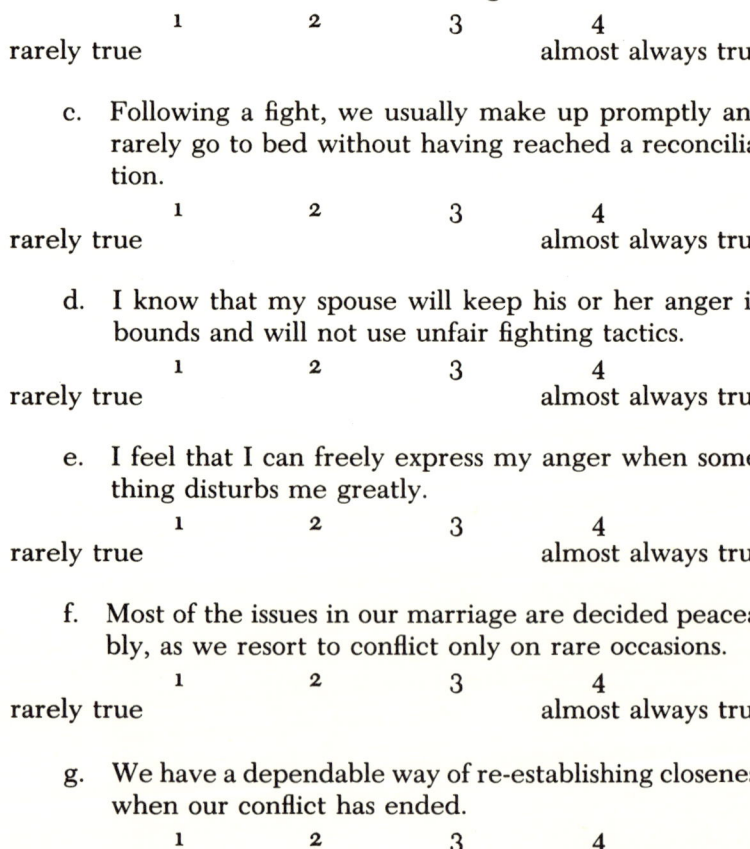

    a.   I feel that my spouse and I have no more than an average number of fights and disagreements.

             1   2   3   4
rarely true   almost always true

    b.   When we have a fight, we generally focus on the issue at hand and avoid name-calling and threats.

             1   2   3   4
rarely true   almost always true

    c.   Following a fight, we usually make up promptly and rarely go to bed without having reached a reconciliation.

             1   2   3   4
rarely true   almost always true

    d.   I know that my spouse will keep his or her anger in bounds and will not use unfair fighting tactics.

             1   2   3   4
rarely true   almost always true

    e.   I feel that I can freely express my anger when something disturbs me greatly.

             1   2   3   4
rarely true   almost always true

    f.   Most of the issues in our marriage are decided peaceably, as we resort to conflict only on rare occasions.

             1   2   3   4
rarely true   almost always true

    g.   We have a dependable way of re-establishing closeness when our conflict has ended.

             1   2   3   4
rarely true   almost always true

Scoring

To get your scores for the marital inventory, add your total score for each cluster of questions. Do this by adding the numbers you have circled. There are seven items in each cluster, each item having a maximum of four points. A top score for each cluster would be 28; a bottom score would be 7.

Interpretation of Scores

Remember: Before you sit down to compare scores, each of you must have written two statements regarding items for which your scoring was 1 or 2:

1. A positive and specific request for a specific change in your spouse.

2. What you are willing to do to assist your spouse in making the changes you have request of her/him.

It is these two statements that you will discuss with your spouse, rather than the numerical score of the inventory.

When Scoring Is Completed

When the scoring is completed, discuss the written statements. In the discussion:

1. Each spouse is to listen carefully and attentively to what the other has to say.

2. The listener is to repeat what s/he understands the other to be requesting and how the other is willing to help. If there is any misunderstanding, repeat the procedure until both spouses understand everything the same way.

Using the Inventory

For couples taking the marital inventory for the first time, no prior scores are available for comparison. However, if the score for a cluster is 21 or less, it indicates that improvement is necessary in that particular area. Lower scores should be viewed only as signals for change, not as signposts on the road to marital disaster. If the current relationship is a difficult one, the couples should, indeed, be elated by some low scores. They have, by their own analysis, discovered specific behavioral areas which when reformed by small changes will improve the marriage by stimulating large improvements in satisfaction.

The inventory you have made is not valid as a rigid means of overall marital assessment. It is valid only as a means of determining the *need for change in specific areas*.

Today's assignment includes *Intimate Time,* which follows.

Intimate Time

Intimate Time is a daily fifteen-minute ritual during which wife and husband sit down together for a private and intimate talk. Scheduled every day throughout the program, it usually is enjoyable. Intimate Time does not have to be the same fifteen minutes every day, but it should be scheduled daily for a specific place and hour. Many couples prefer having their Intimate Time late in the evening, shortly before going to bed.

If there are children, make sure that the circumstances are such that they do not feel the need to interrupt. Children will be included in some of the exercises, but they do not participate in Intimate Time.

During Intimate Time, eliminate every possible distraction, even if it becomes necessary to take the telephone off the hook. Turn off the radio and television. Lock the front door and don't answer the bell.

Flip a coin at the beginning of each session to see who starts the conversation. Begin the dialogue with, "The most important thing which happened to me today was. . . ."

The "most important thing" can be any event which occurred duing the day, regardless of subject. Each spouse discusses her/his "most important thing" for not more than five minutes each.

The last five minutes of Intimate Time concludes with a short sentence by each spouse which starts, "One thing I particularly liked about you today was. . . ." Some examples:
"One thing I particularly liked about you today was the way you smiled at me during breakfast."
"One thing I particularly liked about you today was the way you comforted Mary when she fell off her bicycle and scraped her knee."
"One thing I particularly liked about you today was the way you had the courage to tell Sam and Janet that we couldn't attend their dinner party because we wanted to be home alone this evening."

The requirement is that each tell the other *one* thing which "I particularly liked about you today." If you have several things "you particularly liked," good! The more the better.
That's all there is to Intimate Time. Start the ritual today.

Note: At this stage, most couples want to rush ahead into the next assignment. Restrain yourselves. Do not start Assignment 2 until tomorrow. We repeat: The book is carefully organized to provide the most effective learning pace. Remember: You may invest *more* than one day on an assignment. But do not complete more than one assignment per day.

Assignment 2

In addition to practical exercises, the program contains a discussion of marital theory designed to explain the rationale of the exercises. Also, it will make it clear that a satisfying relationship is not a random affair but rather a series of orderly, cohesive processes which can be controlled by the partners. That is, spouses have the choice of having a good or bad marriage. In today's assignment you will study and discuss "How Your Marriage Got the Way It Is" and "Factors Which Impede Most Couples from Improving Their Relationship." Also, in this assignment we will define what a good relationship is.

THE ASSIGNMENTS

1. Read and discuss "How Your Marriage Got the Way It Is."

2. Read and discuss "The Four Marital Contracts Couples Usually Don't Know They Make."

3. Read and discuss "Factors Which Impede Most Couples from Improving Their Relationship."

4. Read and discuss "The Definition of a Good Relationship."

5. Continue carrying out your daily Intimate Time ritual.

How Your Marriage Got the Way It Is

The substance of marriage has two ingredients: (1) the way the partners behave to each other, and (2) the personal feelings which the behavior exchanges produce.

To improve the quality of a marriage *it is imperative that the partners change their behavior patterns.*

To do this intelligently, it helps to know the genesis of the behavior patterns involved. How did their clusters of personal interactions start? How did they develop? In other words, how did the marriage get where it is?

There is no mystery here. All frequently repeated interactions previously have been arranged and accepted by the spouses. *Very little* that happens between wife and husband is random. No frequently repeated behavioral exchanges occur by accident. *The partners decide to have them happen.* The behavioral exchanges have been settled by mutual, unspoken consent.

Even if the partners are unaware that their frequently repeated interactions have been arranged by mutual consent, the results are as binding to both parties as if accomplished by written contract.

It is in this manner that spouses unknowingly (but decisively) direct the quality of the relationship. If the often-repeated behavioral exchanges are dissonant and angry, that's the way the spouses mutually have agreed to shape the relationship. That is true regardless of each one's complaints, unhappiness, or conviction that the situation is solely the fault of the other.

Likewise, if the hour-by-hour, day-by-day interactions are joyful, that's the way the spouses mutually have transacted and agreed that they want their relationship to be.

Marital conditions do not occur haphazardly or at random.

Dr. William H. Clements* conducted important research in this area during the mid-1920s. Concerning himself with the

*William H. Clements, "Marital Interaction and Marital Stability: A Point of View and a Descriptive Comparison of Stable and Unstable Marriages," *Journal of Marriage and Family*, 1976, pp. 29, 697–702.

subjects of affection, understanding, communication, money management, sex, and responsibility, he asked couples to rate the effect of their behaviors upon each other. He found that in both stable and unstable couples, the spouses *were equally aware of the effects of his/her actions upon the other.* The wives and husbands in both good and bad marriages understood the impact on the relationship of their behaviors towards each other.

This means that each accepted the other's behaviors because *they both had agreed on them.* Unconsciously, in most cases, they had made firm contracts concerning their own and the other's behaviors.

Spouses are surprised when told their destructive behavioral exchanges have been agreed upon and voluntarily accepted. They ask questions along the lines of "If Mary [or John] repeatedly scolds me and downgrades me by telling me I'm incompetent—have I given consent to such undesirable behavior?"

Yes, John/Mary, you have consented to the frequent debilitating behavior you are receiving from your spouse. Your passivity and inaction—even your complaints—may well be your strongest form of agreement.

Negative behavioral exchanges in which one spouse seems to be the aggressor and the other the victim almost always have been arranged by previous consent because: *There is no constant aggressor without a willing victim*—even though the victim may not be happy about it.

The victim is willing and thereby has consented to the unpleasantness. There even is a probability that the victim has a need to be a victim; and often sets up the situation so as to provoke the aggressor. The "setting up" frequently is needed to remind the aggressor of her/his agreed-upon role.

Or perhaps the victim is willing to be the victim as payment for being allowed to "get away" with some bad, unwanted behavior of his/her own.

Whatever the circumstances, both victim and aggressor get some sort of needed or desired satisfaction from the negative interaction—because they have in some way agreed upon it.

Whatever the nature of the relationship, be it heavenly or murderous, the quality of the often-repeated behavioral exchanges is the result of contracts made by the spouses. There is a great variety of relationship contracts. In the following pages we will discuss the four most common ones.

The Four Marital Contracts Couples Usually Don't Know They Make

1. THE SELF-DECEPTIVE CONTRACT OR AGREEMENT

This occurs during the courtship before the wedding. During this period there are exchanges of romantic promises. Both woman and man agree that their courting behaviors will continue throughout their married life. This is self-deceptive. It does not happen that way.

2. THE WEDDING CONTRACT OR AGREEMENT

The second agreement made by married couples is the wedding contract. This is a formal, signed contract, usually defined by civil authorities or the church. Even so, it is another form of deception agreement. Both spouses know that it is improbable that they will carry out all the points of the transaction they have sworn to uphold.

3. THE *QUID PRO QUO* ("THIS FOR THAT") CONTRACT OR AGREEMENT

The third form of contract is the *quid pro quo* agreement. It starts soon after the wedding. Although no words define or initiate this fixed behavioral exchange, it, like the others, grows through mutual consent. *Quid pro quo* ("this for that") is when spouses exchange specific behaviors (both good and bad) on a regular basis. These fixed exchanges have not been discussed by the spouses. They just happen—but they happen by unspoken agreement, by accepted precedent.

An example of *quid pro quo:* Both Mary and John enjoy reading the newspaper in the morning. One morning John took the paper to work. He brought it home at night. After dinner, Mary read the paper in the living room. John, without saying any-

thing, began cleaning the kitchen. This became a daily occurrence. A *quid pro quo* had developed by mutual (but undiscussed) consent.

Another example: A wife usually was about twenty minutes
late when she and her husband were going out to dinner or a
party. When she was late, the husband abused her on points to
which she was particularly sensitive but which had nothing to
do with her tardiness. The wife did not object to the abuse—
even though at other times she would have objected vehe-
mently. The husband's abuse took place in the car. When they
reached their destination the abuse stopped and both became
cheerful. However, the same interaction was repeated whenever they went out together and the wife was late.

It was by mutual consent that the wife was late, and it was by
mutual consent that the husband responded with abuse involving subjects which he would not mention at other times.

Another example: During the American occupation of Japan,
after World War II, thousands of American military men married Japanese women—in preference to American women. We
went to Japan to research this phenomenon. One of the people
we interviewed was a U.S. Air Force major at Itazuke Air Base.
The major had been married twice. The first wife had been an
American who, after nine years of a satisfying relationship (according to the major), had died of pneumonia.

We asked, "Whom did you love more—Betty, your first wife,
or Suzuki, your present wife?"

He said he loved them equally but that he felt more comfortable with his Japanese wife.

We asked, "Is that because Suzuki waits on you hand and foot
in the manner of the traditional Japanese wife?"

The major smiled and said: "I tell you, Suzuki doesn't wait on
me hand and foot. She's been educated in America. The fact is,
I wait on her more than I waited on Betty—because I feel more
comfortable with Suzuki."

He went on to explain; "I'm a test pilot. When I come home
at night I'm exhausted, sometimes jittery. Well, with Betty,
when I came home at night, she didn't give me a chance to get
my breath. The minute I got in the house, Betty started telling
me what a bad day she'd had. The moment she saw me she

began complaining and unloading her problems. At that time, when I was pooped myself, her complaining was too much for me. So I escaped it by withdrawing. For the first hour at home I was sort of a hermit, hiding.

We asked him how the unpleasant and often-repeated interaction had gotten started. He said that it had always had been that way, and that neither had tried to change it. He said: *"I knew what would happen as I walked in the door. And Betty knew exactly how I'd react. However, despite the hour or so of donnybrook every evening, we both felt we had a pretty good marriage."*

We then asked about Suzuki, his present wife. The major said, "When I come home at night, Suzuki greets me as though she's glad to see me—no strings attached. She's glad I'm home. I then go upstairs, have a bath, relax in the tub for maybe a half hour. After dressing, I help Suzuki put dinner on the table. After eating we sit around and talk. It's then that Suzuki tells me about her problems. But by now I'm rested and I have the mood and the strength to listen and help." The major smiled and added: "You see, that's why I feel more comfortable with Suzuki, even though I loved Betty very much. Suzuki gives me that decompression time, and I help her with dinner and her problems."

We asked how and when the comfortable routine with Suzuki had started. The major said it had begun spontaneously several days after the wedding. Suzuki greeted him cheerfully when he came home; and he, after relaxing and eating, listened attentively to her problems and tried to help.

In the above example. the major had had a negative *quid pro quo* with Betty and had a positive one with Suzuki.

Both of the *quid pro quos* had developed by mutual (but undiscussed) consent. The parties involved had contracted to interact that way.

4. THE FAIR-EXCHANGE CONTRACT

The fourth major form of mutual consent to a frequently repeated interaction is brought about by spoken agreement. Called the fair-exchange Contract, it is a *quid pro quo* developed after discussion.

John might have said, "Mary, I'd like to read the paper on the bus on the way to work."

Mary might have replied: "That's okay with me, providing I can read it after dinner. Perhaps you'd like to tidy the kitchen while I read the news."

John might have replied, "Good idea. Let's do it that way."

Mary and John had agreed on a Fair-Exchange Contract.

Summary

There are no frequently repeated behavioral exchanges which occur haphazardly or at random. An interactional pattern which continues over time *never* is an accident. Both spouses have agreed, either tacitly or overtly, that that's how it will be. They accept the situation even though they may complain or be unhappy. When the Air Force major was met every night by a complaining wife, he did nothing to change the situation. He accepted Betty's behavior even though he did not like it. He had agreed to it. When he withdrew and sulked for part of the evening, Betty accepted this even though she probably did not like it. She had agreed to it.

It appears strange that many couples who want to improve their relationship are unable to do so. They both may want to change for the better, yet, despite their intentions, they are unable to *make new contracts.*

There are five major factors which impede people from *making new behavioral contracts* and thus improving their relationship.

Factors Which Impede Most Spouses from Improving Their Relationship

1. THE INABILITY TO VISUALIZE A BETTER RELATIONSHIP

To move ahead to a better relationship, spouses must imagine and specifically define what the new and happier relationship will be like. Where there is an inability to visualize and define, then: (a) spouses do not know in which direction to move, and (b) they may choose goals-for-change which are inappropriate or impossible.

2. LACK OF SKILL

Spouses may have the courage, the vision, and the necessary singleness of purpose for improving their relationship. Yet they frequently do not know how to go about creating the desired marital changes. Unfortunately, a large number of discords often result from fighting over how to go about improving the relationship.

3. ANXIETY ABOUT CHANGE

The prospect of change of any situation—even for the better —almost always produces some anxiety. People generally feel that if they try to change circumstances, they risk failure. Even if they believe they will succeed, still they fantasize and are suspicious. They are apprehensive over possible dissatisfactions in the new and still-unknown environment. Most people are more comfortable and secure in misery than in a proposed new and better environment which they cannot absolutely predict. It is because of this that the program put forth in this book is organized in small, gradual steps—so that behavioral changes occur incrementally and therefore do not incite fear or negative projections.

4. PRIDE

Frequently, both spouses feel that they have been victimized in the marriage. Each expects the other to initiate compensation for that which has been suffered. Therefore, each hesitates to be the first to offer small favors or positive steps which might improve the relationship. There is a reluctance to exhibit the "I will change first" attitude, and a reluctance to boldly take the "I will take positive risks" attitude. Spouses often feel this is "giving in," which might imply that they have been stupid or inferior in the past. Also, they believe that the other spouse might interpret the new generosity as a lack of stability or perhaps weakness. They are frightened lest the other spouse respond with arrogance or increasing demands.

5. INERTIA: AN INHERENT RESISTANCE TO CHANGE

Shortly after the wedding, a "system" develops in the relationship. The system is the sum of the spouses' frequently repeated behavioral interactions. Once this pattern is established, it tends to stay in a steady state. That is, every system resists change. Without knowing it, all members of the system (children as well as spouses) exhibit a natural resistance to changing the system. (For more information on the systems theory, see Part II.)

The Definition of a Good Relationship

There are several characteristics common to most good marriages:

1. Both partners feel they are getting most of what they want from the relationship.

2. Both partners are productive in ways which are important to them as individuals. Some experience individual productiveness through professional careers or hobbies. Others experience it by being in charge of the home and family.

3. Both partners are comfortable about sharing their satisfactions with one another, now as well as in times to come.

4. Both partners are comfortable about sharing the tasks and trials which are unpleasant to the other.

5. Both partners get satisfaction from being supportive to each other.

6. Both spouses have the ability and willingness to adapt to the never-ending changes and circumstances which affect their relationship.

Contrary to what is said in many marriage books, there is an endless variety of paths leading to the achievement of the above goals. Some of the ways the goals manifest themselves to spouses may appear bizarre, even absurd, to other people.

However, as long as the marriage serves the spouses' mutual and individual needs, what others think about it is irrelevent.

The definition of the good relationship deserves further exploring and restating.

In the good marriage both spouses have a high sense of satisfaction from the relationship, an awareness of emotional security, a feeling of zest from coping with the endless problems which are part of life, and the sensibility that they are different but equal.

The spouses are conscious that their good marriage results from the way they treat each other. They know that it is the quality of the exchanged behaviors which establishes the tone of the relationship.

Please note that our description of the good marriage does not include the word love. Love, as we define it, is a result rather than a cause of good marriages.

In our culture, love has many meanings. We love our parents, our children, and one another. We love baseball or tennis or swimming. We love our comfortable old walking shoes or our new car. We love the way our favorite motion picture or TV stars smile.

We love our dog and cat or horse. We love fresh asparagus, new potatoes, and Mom's apple pie. We love to play our guitar and we love to go to the theater. We love the cheerful mailman. We love our doctor as well as the star bowler on our team. We love anything and anyone who pleases us.

The meanings of "love" seem to be endless. For people who are courting (and also those in bad marriages who fantasize about the perfect relationship), "love" is considered a sure means of achieving health, wealth, happiness. "Love" is looked at as an alchemy which will make all dreams come true, will defeat all obstacles, will turn all fantasies into realities. "Love" is considered a holy, indestructible entity. It is sought after as a solve-all gospel. "Love" is supposed to be the sorcerer's wand which turns every sow's ear into a silk purse and every frog into a handsome prince or a beautiful princess, which transforms every dung heap into a mound of roses.

All these meanings of "love" have become accepted through common usage; and almost all dictionaries include them. However, for this book, which concerns itself with marriage and intimate long-range relationships, we must have a more precise definition. We use a version of Harry Stack Sullivan's definition:

When the security and well-being of another person becomes as significant as (not more significant than) *your own security and well-being, you love that person.*

Love, as we define it, seldom is spontaneous, instant, dynamic. It usually takes considerable time to create. It results from work, from thinking, from promoting equality, from being able to cope and to adapt.

Love is not the cause of marital success. Love is the consequence of a worked-for harmony. When you both achieve the six characteristics listed at the beginning of this chapter, and you maintain them for several years, then you probably will truly "love" each other.

Assignment 3

Almost everything in this book concerns the behaviors ex-changed between marital partners. Therefore, it is appro-priate that you understand precisely what behavior is.

THE ASSIGNMENTS

1. Read and discuss "Behavior: What It Is."

2. Read and discuss "Brief Definitions of the Three Major Categories of Behavior."

3. Read and discuss: "A Detailed Definition of Molecular Behaviors."

4. Carry out a five-minute exercise on molecular behaviors.

5. Continue carrying out your daily Intimate Time ritual.

Behavior: What It Is

What is behavior?

Behavior is every observable action we do.

Sitting motionlessly is a behavior. Having a temper tantrum is a behavior. Sleeping is a behavior. Smiling, frowning, smoking, washing dishes, raking leaves, grunting, flirting, not answering a spouse's questions, having sex—all these are behaviors.

Every observable aspect of our lives is a behavior.

In one day we do thousands of things, and therefore we have thousands of behaviors. We walk upstairs, raise our eyebrows in surprise, sweep the floor, walk the dog, purse our lips when our spouses makes a remark, walk away from our spouse, wink, scream, turn our head—all these are behaviors.

It is impossible not to behave.

It is from a person's repeated behaviors that she or he is known to others. From other people's points of view, *we are what we do.* We are our behavior. So it follows that the way we behave determines the treatment we receive from other people.

It is from our spouse's behaviors that we decided whether s/he likes or hates us, respects us, or is contemptuous of us. Therefore, our spouse's behaviors largely determine the way we treat him/her—vice versa.

However, *because we can change our behaviors,* we can change the opinions others have of us, and in that way we can change the way they treat us. Their behaviors towards us improve as a reaction to our positive changes.

The Cherishing Days exercise (Assignments 5 and 6) is the first step in starting this process.

A Summary of the Nature of Behavior

Our individual behavior is every perceivable thing we do, and it is the quality of our own behavior which largely influences the way others treat us.

To improve the satisfactions in marriage, we must change our behavior patterns. That is, we must increase the number of satisfying behaviors which we exchange with our spouses. This includes everything we do—what we speak, the way we speak it, our movements, our gestures, and so on.

Brief Definitions of The Three Major Categories of Behaviors

Because the inventory of marital behaviors is vast, the number of behaviors which we must change (to improve our relationship) is enormous and varied. Therefore, to ensure that behavioral changes can be made in small steps and in an orderly manner, we have divided relationship behaviors into three arbitrary categories. During the program you will work with one category at a time—starting with the simplest and easiest.

At this point we will mention the three categories only briefly:

CATEGORY ONE

These are the small, transitory behaviors such as a cheerful greeting, a blank stare, a brief touch in passing, an approving nod of the head, a frown, a small favor done, a small request neglected. These are the *molecular behaviors. They determine the tone of the relationship.* The molecular behaviors are the ones on which you will make new short-term contracts during the Cherishing Days exercise.

CATEGORY TWO

These are the somewhat larger-scale behaviors which concern the way spouses perform chores and pleasures together. They involve such frequently repeated behaviors as mowing the lawn, going to the theater, painting a room, cooking a meal, going on a trip, taking out the garbage, changing a flat tire, playing tennis, fixing a leaky faucet, making the beds, listening to a concert, having sex. These are the *performance behaviors;* on a day-to-day basis they *determine the way spouses think and feel about each other.* In Assignments 18–21 you will work with the Performance Behaviors.

CATEGORY THREE

Last come the major behaviors. These largely concern making important decisions such as buying a house, deciding where to send the children to school, deciding on what jobs the spouses will have outside the home, where they will live, when and how to have sex. In general, the major behaviors involve decision-making concerning *who does what and when and how* within the relationship.

These are called "authority behaviors." They indicate who has the "power" in the marriage—who is dominant, who has the authority in the various areas of the relationship. Most of the conflicts (the fights-to-the-finish) in a marriage originate around these behaviors. Later, in Assignments 29–34, you will work on the Authority Behaviors and learn how to handle the "power" in your marriage.

At this stage in the program we will concentrate on the "small ones"—the "molecular behaviors." They are the simplest and the easiest to start with, and are the substance of the next few assignments. Therefore, we will explore the molecular behaviors in detail at this time.

The Molecular Behaviors

Spouses respond to each other's frequently repeated behaviors in predictable ways. This is particularly true for the hundreds of brief molecular behaviors which occur daily. At any given moment John knows what will happen if he smiles or frowns in a certain way at Mary. Mary knows precisely what will happen if she smiles or frowns in a certain way at John.

Examples:

Mary nibbles her cuticle. John tenses because experience has taught him that a cuticle-nibble is a sign that Mary is angry about something. John is aware that he has done something which has offended her, and that unless he finds out what Mary is angry about, and the anger is resolved, Mary may behave negatively.

Or, when John places his palm gently on Mary's face shortly after John and Mary go to bed, Mary knows that John loves her very much at that particular moment, and that he is tender and concerned about her.

The tone of the marriage is catalyzed by such small, brief behaviors. These are the molecules of the relationship. The molecular behaviors are the small signals which say "Come closer" or "Keep your distance"; "I like you at this moment" or "I dislike you and distrust you right now."

These small, fleeting behaviors are the red and green traffic signals which govern the relationship from moment to moment. They provide dependable, unceasing cues and stimuli as to when to approach and when to withdraw; when to take over and when to console.

These second-to-second behavioral exchanges clearly and unceasingly display how each spouse *feels* about the relationship, and thus they influence everything else in the relationship.

Individually the molecular behaviors often appear inconsequential. However, they are ever-present, and they come in endless succession. If a preponderance of them is negative, they can destroy a relationship. If a preponderance is positive, they can be the nourishment which feeds the entire marital structure.

If you can control the quality of the molecular behaviors, then anything you desire is possible in the marriage. One of the paths for controling their quality depends upon the Law of Expectations (the subject of Assignment 4).

A Short Exercise

The function of this five-minute exercise is to assist spouses in recognizing each other's molecular behaviors. What you are doing now with these brief behaviors is only the appetizer. The main feast will start in a few days.

Each spouse make a list of three molecular behaviors which s/he has recently observed in his/her spouse. Limit the behaviors to those you recall as being pleasant and desirable. Then, briefly add why you consider them pleasant and desirable.

After the lists have been written, exchange them. Discuss them if you feel like it, and if any other brief pleasant behaviors are recalled, add them to the lists.

Assignment 4

This assignment deals with the Law of Expectations. It is one of the most influential precepts of human behavior, and it is of special influence in intimate, long-range relationships. The Law of Expectations applies to your attitude about "how things will turn out." It is manifested largely through molecular behaviors.

During your marital improvement program, the way you use the Law of Expectations will mold your responses to most of the assignments. More than that, your approach to the Law of Expectations possibly will steer the general direction of your life, both in and outside your marriage. It will have a large influence on your success or failure in solving problems of all kinds.

THE ASSIGNMENTS

1. Read and discuss "The Law of Expectations."

2. Carry out the ten-minute exercise on the Law of Expectations.

3. Continue carrying out your Intimate Time ritual.

The Law of Expectations

The Law of Expectations is both clear and simple: The way you *think* about a situation (the way you expect it to turn out) will determine how you behave vis-à-vis that situation.

As the sociologist George Kelly defined the law of expectations, "a person's processes are psychologically channelized by the way he anticipates events."

If, for example, you expect that an event will turn out well for you, you probably will behave with certitude and optimism. Even your facial expressions, stature, and speech will reflect your attitude of positive expectation.

Other people who are involved in the situation will perceive your confidence. They will notice that you are behaving as if the event will be successfully concluded, and that you appear willing to take positive risks. It is natural for the others to reflect your attitude and therefore be willing to cooperate. Therefore (because of the way you think) the event has a greater chance of turning out the way you want it to.

On the other hand, if you expect an event to be a disaster, you may behave with discomfort, awkwardness, and negativeness. You are behaving as if you may fail. You will be reluctant to take positive risks. Other people who are involved will react similarly. Thus your negative expectations may push the event toward failure.

An anecdote will illustrate:

Suppose you are about to be examined for your driver's license. Your expectations are that you will have problems taking the test. You think that you probably will fail it.

As a result of your expecting difficulty and possible failure, you may become tense, perspiring, nervous, tentative. Your negative behavior will be noticed by the officer conducting the examination. He then may *think*, "This person is not well prepared." Therefore, his behavior will be influenced. He probably will observe your driving more closely, and will be especially alert for your errors.

You will notice the officer's behavior—his/her terse manner, his/her extra strictness; and you may *think*, "S/he's in a bad mood. S/he's got it in for me!" This increases your expectancy of failing. Your nervousness increases. Your physical coordination becomes more awkward.

In this example, your negative expectations—the pessimistic way you *think*—influence your behavior to be negative, which, in turn, pushes the event in the direction of failure.*

Contrary to this, if you expect your driver's license test to turn out well, there is a high probability that you will be relaxed and confident. You behave "as if" you are a success. When decision have to be made, you are inclined to take positive risks. Your positive behavior will suggest to the examining officer that you are competent and experienced. You, in turn, will observe the officer's easygoing manner. Your confidence increases, as do your physical coordination and competence.

When you have *positive* expectations, your behavior assists the event to turn out successfully.

There are, of course, ways, to transform negative expectations into positive ones. In the case of the driver's test, prior

---

*A. T. W. Simeons, M.D., in *Man's Presumptuous Brain* (New York: E. P. Dutton, 1961), provides a physiological theory which explains the damaging influence of negative thinking. As condensed and paraphrased, it states:

The brain stem—which is the first-developed part of the brain—is an instinctive working organ. Therefore it is unable to process information which it receives. It only reacts directly to the information. When the brain stem receives a signal of danger, it automatically stimulates the body to fight or flee, or to perform whatever survival behavior is appropriate.

The signals which the brain stem receives come from the cortex, which is the part of the brain which developed later. The cortex can reason, censor, take incoming signals and decide how to handle them, and decide whether they are dangerous or desirable. The cortex will do this if we allow it to. Using the cortex is what is known as being conscious. It is possible for us to control the cortex —unlike the brain stem, which simply responds and is unable to censor or determine whether an apparent danger signals represents an authentic danger.

In the example just given in the text above, you have the thought, "I'll probably fail my driver's examination." If such thoughts persist, the brain stem may get a message from the cortex which says, "There is a danger of failure." The brain stem reacts as if the signal indicates a real, threatening danger— because the brain stem is unable to assess or analyze the message. It therefore sends danger signals throughout the body. As a result, your pulse increases, your breathing becomes faster and perhaps shallow, your muscles tense. You exhibit signs of fear and perhaps hostility and aggression.

studying and prior rehearsing would make you confident of your knowledge and skill. As for your marital relationship, this book's program leads you through necessary studies and appropriate rehearsals so that you become confident that you can make the relationship supply high satisfactions for both of you.

Let us return to negative expectations.

In marital relations, many spouses are impeded by negative expectations (thoughts) over which they have no control; and their pessimistic, suspicious thinking usually is projected into observable behaviors—a frown, muscle-tensing, an abrupt silence, the turning away of eyes or body, a hesitancy in speech, and so forth. These negative thoughts and their resulting physical projections often are conditioned reflexes, old habits acquired and accumulated over many years. Whether or not they are done ordinarily and unconsciously is irrelevant. The fact is the other spouse observes them, interprets them, and usually reacts negatively; and a quarreling behavior cycle is initiated.

Unless these undesirable negative thoughts are controlled or transmuted, there is bound to be discord which is destructive to the marriage.

It is reasonable to suggest, "If negative thoughts and their negative expectancies are so destructive, well, just get rid of them!"

That's a good idea, but it is very difficult to control the kaleidoscope of thoughts which flashes through our minds. The thoughts appears whether or not we want them; and when they are negative, they have a deteriorating effect upon our relationship with our spouses. We appear to be helpless.

We are behaving like unconscious (and destructive) puppets controlled by the formless past.

To control the negative mental images and their negative projections, spouses consciously must learn to signal each other: "Even though we disagree and I appear annoyed, I approve and am supportive of you. I like you. I am committed to the relationship and I treasure it."

The Cherishing Days exercise, which begins in Assignment 5 will demonstrate the process which makes this possible. In that exercise, spouses list positive molecular behaviors they would like to receive from each other. Then they voluntarily exchange them. Thus, they are behaving "as if" the relationship is a successful one—even though in reality it may be rocky at the moment.

When you *behave* in the "as if" manner (despite your negative thoughts) you are sending messages of positive expectations, and it is probable that your spouse will respond—at least a little—in the same way.

When this atmosphere of positive expectations is initiated, it provides an environment in which it is possible to negotiate and solve the specific problems of the relationship. At this stage, we are creating the desirable environment. Later, you will learn how to identify the problems and negotiate them.

When the problems are negotiated and thereby reduced, then the negative mental images and suspicions gradually will be diluted; and in time they will be easy to cope with.

Years ago some psychiatrists and psychologists believed that "as if" behavior resulted in a dangerous suppression of negative thoughts. There is impressive evidence now that they were wrong, and that quite the opposite happens.

Behaving "as if" forces you to become aware of your negative thoughts. This is an effective way of "letting loose," of letting them out of the box, and thus unchaining yourself from the tyranny of old negative thoughts and habits.

When you have negative thoughts and still behave "as if you like and respect your spouse," you are for all practical purposes transmuting negative energies into positive and useful ones. Such is the power of positive expectations.

Another aspect of the Law of Positive Expectations is the "I will change first" principle. This involves your initiating small cherishing behaviors—not waiting until your spouse starts the process. This is an important variation of "as if." When you initiate the cherishing behavior—especially in times of difficulty

—you are showing your spouse that you are contributing caring and supportive behavior before you receive the same. This gives your spouse a feeling that s/he is valued and trusted, and that you treasure the relationship enough to take positive risks.

The vitality of the Law of Positive Expectations comes from taking positive risks.

A Ten-Minute Exercise

1. Each spouse define the Law of Expectations. Flip a coin to see who starts off.

2. Each spouse give an example (making it up if necessary) of the Law of *Positive* Expectations in action within your own relationship.

Assignment 5

This is the beginning of your Cherishing Days exercise. No matter how difficult your relationship is, we are certain your spirits will be lifted within the next several days. The exercise is easy. It is joyful. It requires no long-range commitments because the exercise lasts only five days.

Today's assignment is the day of preparation. You will make a list of about ten small, pleasing behaviors which you would like your spouse to give you. S/he will make a similar list. In tomorrow's assignment you will give these behaviors to each other.

If you have a relationship which at present is heavy with quarrels and discords, you may question whether the Cherishing Days exercise is possible for you and your spouse. The answer is yes, as you soon will find out.

THE ASSIGNMENTS

1. Read the instructions and then make preparations for the Cherishing Days exercise.

2. Continue carrying out your Intimate Time ritual.

Preparing for the Cherishing Days Exercise

For this exercise each spouse will make a list of small cherishing behaviors which s/he would like to receive from the other spouse. Spouses will exchange these cherishing behaviors for five days, commencing tomorrow. These requested small behaviors must have four characteristics, as follows:

1. They must be specific and positive.
For example:

John would like Mary "to sit next to him on the sofa when they listen to the news after dinner." This is positive and specific, which is different from asking her to "stop being too preoccupied and distant" (a negative and overly general request).

Mary would like John "to kiss her goodbye when they part in the morning." This is positive and specific, which is different from "stop being so distant and cold" (a negative and overly general request).

2. The small cherishing behaviors *must not concern past conflicts*. Your requests must not be old demands. That is, the requests must not concern any subject over which the spouses have quarreled.

3. The behaviors must be those which can be done on an everyday basis.

4. The behaviors must be minor ones—those which can be done easily.

Using the guidelines just discussed, prepare a list of small cherishing behaviors you want your spouse to do for you for five days beginning tomorrow. Do not be dismayed by the fact that these gestures are not spontaneous. It is the action that is important for now, not the inspiration.

The sample list which follows consists of cherishing behaviors which have frequently been requested by couples taking the course.

A Sample List of Cherishing Behaviors

1. Greet me with a hug and a kiss before we get out of bed in the morning.

2. When you are out walking, bring back a flower or a leaf.

3. Look at me and smile.

4. Call me during the day and tell me something pleasant.

5. Turn off the lights and light a candle when we have dinner.

6. Ask me how I spent my day.

7. Pick me up at the bus stop sometimes as a surprise.

8. Tell me how much you enjoy having breakfast with me.

9. Tell the children (in front of me) what a good parent I am.

10. When we sit together, put your arm around me.

11. When we are together at home, ask me what record I would like to hear, and then play it.

12. Wash my back when I'm in the shower.

13. Have coffee with me in the morning before we wake the children so that we can have a five-minute talk together.

14. Hold me at night just before we go to sleep.

15. Ask my opinion about world affairs after we watch the news.

16. For no special reason, hug me and say you like me.

17. Hold my hand when we walk down the street.

18. When you see me coming up the drive, come out to meet me.

19. Put a surprise note in my lunch bag.

20. When we're together, end your sentences with "dear" or "sweetheart."

21. When we part in the morning, blow a kiss to me.

When you are working on your lists, recall some of the delightful behaviors which drew you to the other during courtship or which were practiced during the happiest times in your relationship; or refer back to the three desirable behaviors you listed in the small exercise in Assignment 3.

Many of the cherishing behaviors which you request of your spouse may seem unimportant or even trivial. Some may be a bit embarrassing because of the seeming unreality of the tenderness in them. That's fine. These small behaviors set the tone of the relationship. They are the primary building-blocks for a satisfying marriage, the molecules of the relationship. They establish an environment of positive expectations.

When the lists are completed, exchange them.

Discuss the cherishing behaviors you have requested of each other. Don't be shy about telling your spouse *how* you would like to have the cherishing behaviors done for you.

For example: "John, remember the way you used to bring me a flower when we were first married? You presented it to me when you met me at the door—after you had kissed me. It made me feel really loved."

During the discussion it is probable that both spouses will think of a few more cherishing behaviors which they'd enjoy receiving. Add them to the lists. The more the better, providing the lists are approximately equal in length.

Tomorrow you will begin the first of the five days of the actual Cherishing Days exercise.

Don't forget to have your Intimate Time ritual today.

Assignment 6

Today you will concentrate on giving each other the requested cherishing behaviors. For some of you who have very discordant relatiionships, the process may seem artificial at first, but by the second day it is highly likely that it will become quite natural. Even spouses who have been fighting bitterly usually are surprised at their ability to exchange these small acts of tenderness, and at how easily they come. The spouses, therefore, experience an increased sense of positive expectancy and begin to feel that an improvement in the relationship is feasible.

THE ASSIGNMENTS

1. Begin the Cherishing Days exercise (the first day of five).

2. Begin keeping records of the cherishing behaviors received.

3. Explaining Cherishing Days to Your children.

4. Continue carrying out your Intimate Time ritual.

Cherishing Days: The First Day of Five

You know how to do this, so go ahead. Start early in the morning and keep at it until you are asleep at night.

Keeping the Records

Put the two lists of requested behaviors in a conspicuous place, say, taped to the front of the refrigerator. When one spouse receives a requested behavior from the other, the receiver puts the date in the margin next to the request which has been granted. For example, assume that the exercise started on March 26, and that on that morning John hugged and kissed Mary before they got up in the morning. This is Request No. 1 on Mary's list. Therefore Mary writes the number 26 in the margin next to Request No. 1.

Whenever a cherishing behavior is received, put the date opposite the appropriate place on the list.

This daily record provides information about how many desirable and positive small behaviors have been generated within the marriage during the five-day exercise. *It assists each spouse in knowing when s/he should give even more.* What matters in Cherishing Days is not an absolute equity in the range of nurturing behaviors which each spouse makes towards the other, but rather the consistency with which both partners exhibit their willingness to make small investments in trying to improve their lives together.

Explaining the Exercise to Your Children

If you have children, they will be quick to notice a small positive change in your behaviors. They will wonder what is happening and it is essential that their curiosity be satisfied. Explain the Cherishing Days exercise to them. The explanation may be difficult; be patient. You are changing their home environment, the system they are used to. They are very influential in changing the tenor of the family system. Again, as we suggested earlier, unless the children participate, cooperate, and understand, the improvement in the family system—which includes the wife-husband relationship—will be slowed considerably.

For most couples this is an exciting day. Don't forget to carry out your Intimate Time ritual.

Assignment 7

On this second day of exchanging cherishing behaviors, most couples experience an upswing in the number of cherishing behaviors they are giving each other. This upswing will continue to increase throughout the five days. When the exercise is continued into the sixth day and beyond, there will be fluctuations in the cherishing behaviors present within the relationship. These fluctuations are quite normal and will be explained in Assignment 9.

THE ASSIGNMENTS

1. Continue the Cherishing Days exercise.

2. Invite your children to join the exercise.

3. Continue carrying out your Intimate Time ritual.

Cherishing Days: The Second Day of Five

You know how to do this, so go ahead. Start early in the morning and keep at it until you are asleep at night.

Bring Your Children into the Act

Ask your children if they would like to join you in the Cherishing Days game. If they do (and most children will, from the age of four up), assist them in making short lists. Their lists should have from three to five cherishing behaviors which each child would like from mother, father, brothers, and sisters. Put their lists in the same prominent place that yours is.

The spouses should make similar lists of what they would like from the children.

Make certain you explain that the cherishing behaviors must be (1) positive and specific, (2) easily accomplished, (3) something which can be done every day starting now, and (4) something which is free from a history of conflict.

Assignment 8

By now the number of cherishing behaviors exchanged should be somewhat higher than it was the first day. You may be experiencing a "high" by now. Taking advantage of this, we will introduce you to a simple but rather profound communication facilitator, the "Why and How" exercise.

THE ASSIGNMENTS

1. Continue the Cherishing Days exercise.

2. Read and discuss "The 'Why and How' Principle."

3. Carry out a short "Why and How" exercise.

4. Continue carrying out your Intimate Time ritual.

Cherishing Days: The Third Day of Five

You know how to do this, so go ahead. Start early in the morning and keep at it until you are asleep at night.

The "Why and How" Principle

Obtaining information from each other is one of the most important elements of the communication process. Later, we will deal with the total subject. However, at this stage of the program, it is appropriate that you learn one special aspect of the asking-questions problems. It is the "Why and How Principle."

It is a daily occurrence for partners to seek information from each other regarding the other's

> motives
> attitudes
> behaviors
> reactions to something requested

When you want this kind of personal information, *beware of questions which begin with "why."*

The reason for this: In our society, "why" questions about personal behavior often imply "You are bad."

However when "how" is used instead of "why"—as will be explained shortly—an opposite meaning is implied. A "how" question tells your spouse, "You are important to me and I value our relationship."
Example:
A wife married to an automobile mechanic asks him: *"Why* do you come late for dinner and in dirty clothes? *Why* do you set bad examples for the children?"

The message which the husband probably hears is: "You are bad. You have no sense of punctuality and are a slob. You don't love your children.

His resentment of the accusation he experiences is not likely to improve his performance.

It would be more effective and constructive communication (that is, the wife would be more likely to get what she wants) if she had said something like: "I know you're working very hard and sometimes have to stay overtime. However, it means a lot to me and to the children to have us all sit down for dinner together and enjoy a relaxed evening. *How* can I help make this possible?"

Note how such a "how" question promotes a dialogue in which the problem is discussed with good will. In this example, wife and husband agreed that husband would telephone if he expected to be late and the wife would delay the meal—or feed the children early. This is a simple commonsense solution. But it might not have been found had the wife started her query with "why."

Another example:
A husband asks his wife, "*Why* does the house always look so dirty?"

He may believe that the house is frequently dirty and may not be able to understand why his wife isn't more efficient. However, asking about it with a "why" question usually is an effort neither to get accurate information nor to improve the neatness of the house; and, certainly, it will not improve the relationship.

The wife probably hears, "Why are you so incompetent and uncaring?"

This destructive question may provoke the wife to anger. It may frighten her. But the chances are that it will not get the house cleaner. In fact, the house may become even dirtier because the wife will resent the question. She will resent having to do all the housework alone; and she will be annoyed because the "why" question defines her as incompetent and slovenly.

If that's the way she's defined, that's probably the way she'll be.

It would be more effective communication (that is, the husband would be more likely to get what he wants) if he had said something like, *"How* can I help you by sharing the difficult job of running the house?"

(Note how such a "how" question promotes a dialogue in which the problem is discussed with good will. In this example the husband and wife agreed that the husband should share the tasks of cleaning the house. This is an obvious commonsense solution, but it is unlikely to have been reached had the husband framed a "why instead of a "how" question.

A Short Exercise

Each spouse make up two negative, personal questions starting with "why."

Ask the questions of each other.

Discuss your feelings, your responses to the two "why" questions. Did you like being asked the questions? Did you feel like responding positively? Were you inclined to give your spouse what s/he asked for?

Change the two "why" questions into "how" questions.

Ask the "how" questions of each other. Answer the "how" questions and go into any dialogue which seems natural.

Discuss your responses and feelings after hearing the more positive "How can I help?" questions.

Assignment 9

Almost all couples who have gone this far in the program are amazed at the lift they have received from the Cherishing Days exercise. They wonder how it is that such a simple procedure has so much power. In this assignment, the secret will be explained.

THE ASSIGNMENTS

1. Continue the Cherishing Days exercise.

2. Read and discuss "The Secret of Cherishing Days."

3. Read and discuss "What Happens After Five Days."

4. Do "A Small Decision-Making Exercise."

5. Continue carrying out your Intimate Time ritual.

Cherishing Days: The Fourth Day of Five

You know how to do this, so go ahead. Start early in the morning and keep at it until you are asleep at night.

The Secret of Cherishing Days

Cherishing Days is an effective exercise with most couples because:

1. The new positive behaviors which the partners exchange are small, *requested,* and agreed-upon. Under these circumstances, it is almost impossible for the new behaviors to backfire. Therefore, the new behaviors do not provoke a fear of the unknown—a fear which is inherent in all of us.

2. The agreed-upon cherishing behaviors result from a temporary (five-day) commitment. This brief commitment period makes it easy for the spouses to take positive risks.

3. Because the cherishing behaviors are molecular, the spouses experience the comfort of progressing by "small steps." There is no fear of embarking on an impossible project.

4. During Cherishing Days, both spouses experience the important I Must Change First principle. Each voluntarily—and independently—gives the other some cherishing behaviors. This promotes both self-confidence and other-confidence, and, as a result of this, spouses anticipate further relationship satisfactions. Because they anticipate even more improvements, *they behave accordingly.* This means they are benefiting from the Positive Expectations principle.

5. During Cherishing Days, the partners behave (at least a little) "as if" they cherish each other. At first this may be awkward and somewhat contrived—a positive experiment. However, as each partner gives and receives more cherishing behaviors, each begins to *feel* better about the other. Even though

their overall relationship still may be in discord, spouses now have an inkling of how much better the marriage possibly can become.

6. For five days almost all spouses receive only pleasure from the cherishing behaviors.

What Happens After Five Days

If the exercise are continued over an extended period beyond five days, the number of cherishing behaviors which are exchanged will tend to fluctuate. There is a definite up-and-down pattern common to almost all couples (see Figure 1). After about the fifth or sixth day the number of cherishing behaviors exchanged diminishes temporarily.

Without being aware of what they are doing, the partners begin to test each other to find out whether the other will continue giving cherishing behaviors unilaterally. In this way, each tends to withdraw from fully giving. Each begins measuring the other's "givingness." Each wants to know whether the improvement in the relationship—and especially in the other spouse—is real.

Figure 1 shows the testing process for most couples who have unhappy relationships when they continue the exercise beyond five days. There are three major testing periods, occurring during the fifth through the eleventh day, the seventeenth through the twentieth day, and the twenty-second through the twenty-third day. This pattern develops for almost all dissatisfied couples.

If the couples want to continue the cherishing days beyond five days, they should—regardless of what their thoughts are during the testing periods—make strenuous efforts to continue *giving* cherishing behaviors to each other. They should encourage each other and remind each other that they are in the process of improving their relationship, and, therefore, that they must avoid such testing pitfalls.

On about the twenty-eighth day will come the consolidation phase. From there on the exchange of cherishing behaviors will,

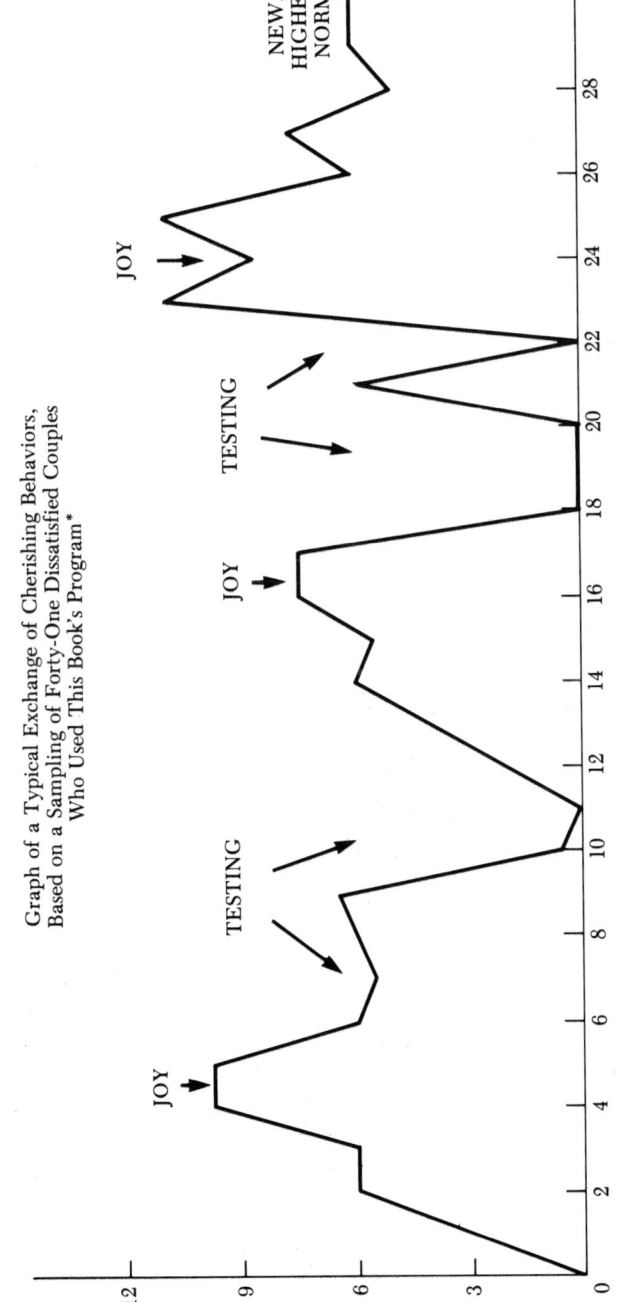

FIGURE 1

Graph of a Typical Exchange of Cherishing Behaviors,
Based on a Sampling of Forty-One Dissatisfied Couples
Who Used This Book's Program*

JOY

TESTING

JOY

TESTING

JOY

NEW,
HIGHER
NORM

*At the Behavior Research Institute, Peacham, Vermont. Similar
testing has been done by Richard Stuart at the University of Utah.

FIGURE 2

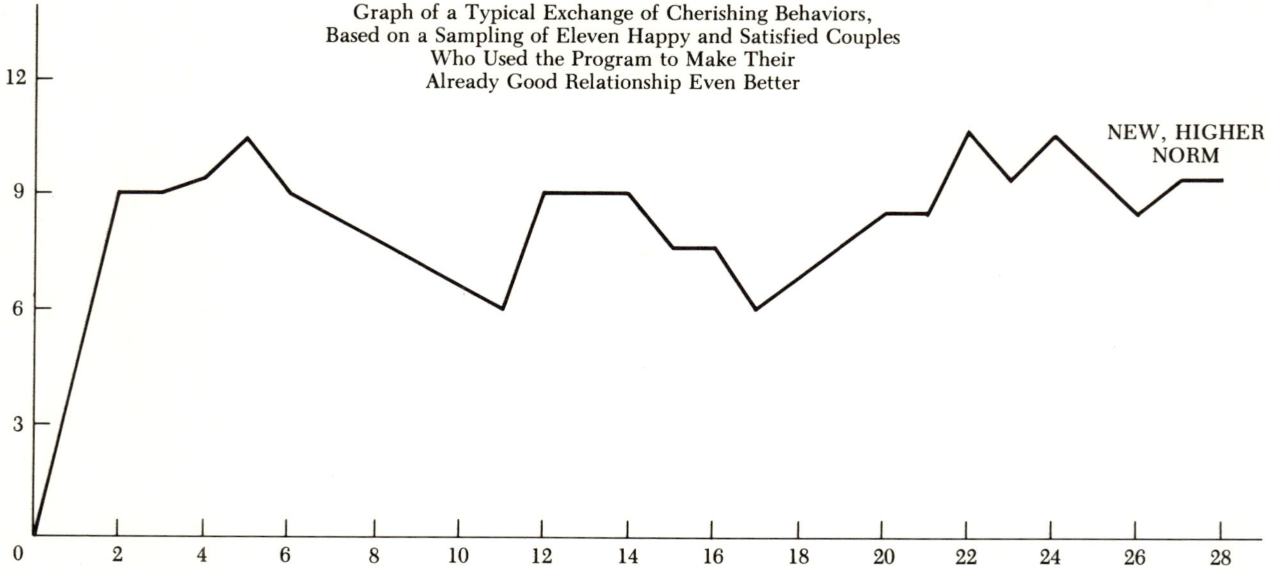

Graph of a Typical Exchange of Cherishing Behaviors,
Based on a Sampling of Eleven Happy and Satisfied Couples
Who Used the Program to Make Their
Already Good Relationship Even Better

regardless of fluctuations, remain on a new average high. The cherishing behavior has begun to become a natural habit. A new norm for the relationship has been established. When couples reach this higher and more satisfying level, they usually try to keep it that way.

We assigned cherishing days for only five days because at the beginning of the program almost all couples badly need a guaranteed lift.

We are confident that most of you have experienced this lift.

Figure 2 shows the graph for happy and satisfied couples who were trying to make their marriages even better. The significant thing here is that because of their already good relationship, they have a stronger feeling of positive expectations. Therefore, even though there are fluctuations in cherishing behaviors exchanged, the "testing" is minimal.

A Small Decision-Making Exercise

Discuss the text you have just studied. Then decide whether both of you want to continue cherishing days every day for the rest of the program—even though they are not assigned.

Assignment 10

This is the last assigned day for Cherishing Days. By now, perhaps, you can tell in advance when your spouse is about to give you a gift of cherishing behavior. You may be able to detect a facial expression or a certain way of standing or moving that preceeds the action or words. Such communication is called "kinesics," or body language. Kinesics may well be the most effective means of communication between spouses. In this assignment, you will begin learning what it is all about.

THE ASSIGNMENTS

1. Continue the Cherishing Days exercise.

2. Read and discuss "Kinesics (Body Language), a Definition and an Explanation."

3. Make a decision regarding Cherishing Days for the future.

4. Continue carrying out your Intimate Time ritual.

Cherishing Days: The Last Day of Five

You know how to do this, so go ahead. Start early in the morning and keep at it until you are asleep at night.

Kinesics (Body Language)

During courtship and the early months of marriage, couples are in the stage of maximum talk. They exchange more words at the beginning of their relationship than they do later. The longer they live together, the more the balance of communication swings away from words and toward body language.

After two people have lived together for several years, a message which once required a paragraph or two of speech is communicated by a gesture, a glance, a silence, a facial expression, or a way of standing or moving. It is only during the last forty years that scientists have systematically analyzed the function and extent of body language as a part of human communication. This study is called kinesics.

In the 1960s, Don D. Jackson, M.D., took silent films of families receiving therapy. The filming was done through a one-way glass. The researchers could not hear the speech of the families. Later the films were played back in slow motion and studied by Jackson and his colleagues. By watching the individual body movements, expressions, and gesticulations it was possible to identify the family's problems quickly and accurately.*

After studying the films of several hundred families, Dr. Jackson and his colleagues suggested the following conclusions:

1. Among family members, especially husband and wife, linguistics *alone* produces only about 15 percent of all communications; the other 85 percent is largely the product of kinesics (body language).

---

*We suggest that kinesic observation has made the old analytic dependency on "listening to the patient on the couch" an obsolete diagnostic tool. Today one must look at the patient (preferably in his/her natural environment), as well as listen.

2. The person sending a message by body language frequently is not aware that this kinesic communication is taking place. Nevertheless, the other person—the message-receiver—reacts to the kinesic message which s/he has observed. *And the way the other person reacts is predictable.*

3. The message-sender frequently has a different perception of his/her behavior than does the observer (the message-receiver). For example, the message-sender may perceive him/herself as being friendly and helpful. The message-receiver, however may interpret the message-sender and his/her message as hostile.

The spoken words may give one specific message, whereas the body language may transmit a conflicting message.

4. In discordant relationships, there is a high frequency of conflict between linguistic and kinesic messages, and this is confusing to the message-receiver.

One example is a recent case history from the Behavior Research Institute:

A psychiatrist came to us seeking assistance in straightening out his own marriage. (We mention that he was a psychiatrist to illustrate that even people who are trained observers and who are skilled in abstract thinking can go amiss on body language).

The psychiatrist and his wife had been married twenty-three years and had four children. Neither wife nor husband wanted a divorce, yet they had found themselves drifting toward one and had already consulted lawyers.

We went to their home and videotaped them during dinner.* Halfway through the meal, Marjory (the wife) said to her husband, "Michael, let's go to the opera on Friday night."

*The evening meal is the period of strongest family ritual—with all members present—and at that time the family system of behavioral exchanges can be observed in its purest form. We strongly feel that therapy should be done in the home with all members present, and that much of the observing should be done at mealtimes.

For a few moments before making this request, Marjory had been moving her wedding ring up and down her finger. As she began talking, she sat up straight and put her right clenched fist into the open palm of her left hand—a kinesic signal which often indicates an attempt to gain dominance.

"What?" said Michael, running his right forefinger across his nose and momentarily looking away from Marjory—a kinesic signal which in our culture usually indicates that the listener believes the speaker has said something which is inappropriate.

"Let's go to the opera on Friday night."

Michael moved his hands forward in the cupped position—a kinesic signal often used when telling a story or indulging in fantasy. He looked directly at Marjory, eye to eye, and said, "Why Margery, I'd *love* to take you to the opera on Friday night."

"Thank you," said Marjory, staring at her plate and rubbing her nose. "Does anyone want more roast beef?"

This behavioral exchange—composed of both linguistic and body-language messages—lasted twenty-one seconds. After this, the meal continued. Other subjects were discussed by the wife, the husband, and their four children.

At the end of dinner, the family gathered around the TV monitor to watch the playback of the videotape.

We rolled the tape to the twenty-one-second episode between Marjory and Michael. Before showing it, we asked Michael, "Michael, did Marjory make a request of you during dinner?"

"Yes."

"What was it?"

"She asked me if I'd take her to the opera on Friday night," he said, smiling smugly, almost victoriously.

"How did you handle her request?"

"I told her I'd *love* to take her to the opera. I gave her exactly what she requested."

"Watch this," we said, playing back the episode. Then we quickly played it back twice more.

Michael looked down, wringing his hands. Then he put his hands on his head, rubbing it in a manner indicating deep distress.

"My God," he said. "I told Marjory I'd love to take her to the opera. I said I'd *love* to take her—but my face was twitching with dislike."

Their twelve-year-old daughter interrupted, "Daddy, that's the way you always look when Mom asks you to do something for her."

Michael looked around the table. The other three children nodded.

A very emotional and touching scene followed. With tears running down his cheeks, Michael went to Marjory, hugged her, and begged her forgiveness. She, nodding her head and hugging him back, began weeping.

When she had dried her tears, she took Michael's hand, kissed it, and said, "Michael, I knew that Friday is your tennis night and that you dislike the opera. I'm not blameless either."

Note: By watching themselves on videotape, it was possible for Marjory and Michael to recognize the reality of their behavioral exchanges, and for Michael in particular to perceive himself as others experienced him; and to observe the conflict between the spoken messages and the kinesic messages.

The video recorder is a helpful tool; however, few people have access to this expensive equipment. Therefore, we created the Body Language exercise, which in some ways is superior to videotape in learning about kinesics within the relationship. You will learn about it in the next assignment.

The Future of Cherishing Days

Today is your last Cherishing Days assignment. If you have decided to continue Cherishing Days on a voluntary basis (good idea!), then add a few more requests to your cherishing behavior lists.

Remember that tomorrow—which will be the sixth day—is when most couples begin "testing and exploring." If you can stop the "testing and exploring" (that is, do not allow yourselves to reduce the number of cherishing behaviors you exchange), you will have accomplished much, *very much indeed!*

Assignment 11

We urge you to have your children participate in this exercise. They will expose you to kinesics which neither of you have known about. Children are more aware of body language than are adults, and often are more competent at mimicking it.

Some of the things you will do in this exercise may at first seem novel, and you may be shy. But particularly if the children participate, the exercise can become a wonderful and merry game.

THE ASSIGNMENTS

1. A Body Language exercise.

2. Do "An Additional Exercise: A Discussion."

3. Continue carrying out your Intimate Time ritual.

4. Continue Cherishing Days with new cherishing behaviors added to the lists (if you have agreed to do this).

A Body Language Exercise Concerning Your Spouse's Positive Messages

To repeat and emphasize,

Perhaps as much as 85 percent of all messages between spouses are transmitted at least in part via kinesics, not by speech alone.

The message-sender who signals via body language often is not aware that s/he has done so. S/he may not be conscious of his/her gestures, expressions, body movements and so forth. S/he may not be aware of them *even though they usually are projections of the message-sender's true feelings or thoughts.*

It frequently happens—especially in discordant relationships —that the words coming from the mouth give a message of approval or permission; but simultaneously, body language sends a conflicting message of disapproval or non-permission. When this occurs, the message-receiver *hears* the spoken message but at the same time *sees* a conflicting message in the body language signal.

This confuses the message-receiver. S/he does not know which is the truly felt message.

The message-receiver may ignore the spoken message and respond only to the body-language signal. The message-sender usually then becomes equally confused, and communication breaks down.

It is in this way that many unintended discords and conflicts are introduced into the relationship. The Body Language exercises are designed to diminish confusion and promote clarity. The exercises can assist you to:

1. Become aware of the body language messages which *you* constantly send your spouse.

2. Become skilled in correctly interpreting the body language messages which your spouse constantly sends you.

3. Have both spouses develop a fluency in their non-spoken dialogues.

4. Learn how to bring the spoken communication and the body language communication into harmony.

The first part of the exercise is designed to assist you in becoming aware of your spouse's body language—to recognize that when s/he raises an eyebrow, crosses arms, turns away, inclines his/her head, and so forth, that s/he is sending you a meaningful message.

The First Exercise

This exercise concerns the body language for positive behaviors only. You will be describing and pantomiming movements, expressions, and gestures which give positive messages. (In Assignment 12 you will do the same for negative messages.)

1. Describe in writing what your spouse does (not says) when s/he is telling you by body language that s/he cares for you, loves you.

Example: "When John cocks his head to the right, smiles, crinkles his eyes, and looks directly at me, I have the feeling he loves and cares for me."

2. Describe your spouse's body language when it tells you s/he approves and respects something you are doing or intend to do.

Example: "When Mary looks me in the eye, nods her head, and places her hands on her hips, I feel she approves and respects what I am doing."

3. Describe your spouse's body language when it tells you that s/he will help you solve a problem or assist you in a task.

Example: "It seems to me that when Mary goes into the kitchen, puts the coffee on, claps her hands once, and points to the chairs—indicating that we should sit down and talk things over—she is sending a body language message that she will help me solve a problem.

Another example: "It seems to me that when John runs his hand through his hair, then reaches out, squeezes my shoulder gently, and smiles, he is sending me a body language meassage that he will help me with a problem."

Spouses Can Help Each Other with the Body Language Exercise

After the three descriptions have been completed and written (even if you are not satisfied with them), you can help each other make the written descriptions even better.

Each spouse act out (for the other to watch) how s/he believes s/he behaves when:

1. S/he wants the other spouse to know—via body language —that s/he cares for and loves the other. It can be a squeeze, a wink, a smile, a way of standing a caress.

2. S/he wants the other spouse to know s/he respects and approves what s/he is saying or doing.

3. S/he will help the other solve a problem.

Doing this bit of acting of positive behavior with the other spouse, watching may appear a bit awkward*—especially if the spouses have been fighting a lot—but, again, do the best you can. Some spouses are shy about this and may caricature their acting. That's all right too. The basic kinesics still will be present; and the goodwill effort involved will be nourishing.

After each spouse has acted out three emotions for the other, spend the rest of the time (until the next assignment) observing each other closely. Try to identify any and all looks, motions, postures which indicate that your spouse (1) cares for you and loves you, (2) respects what you are doing or saying, and (3) is willing to help you solve a problem or task.

*It is ironic that most people find it easier to act out their negative emotions than they do their positive emotions.

Before starting the next assignment, make your final drafts of the three descriptions. If you feel you need another day or two to be comfortable about this, take the extra days.

An Additional Exercise: A Discussion

Read the three descriptions aloud to each other.

Discuss them. Is each of you aware that you are behaving that way? Is there any way either of you can change the body language a little so that the meaning is clearer or more pleasing?

While discussing the descriptions of the three positive kinesic messages, you probably will recall others. Discuss them.

For example, John might say, "Mary, when you wrinkle your nose and smile, I know you're telling me you're pleased with what I've just done or said."

Or Mary might say, "John, when I ask you to do something for me and I see you pull on your ear for a moment, it means you're trying to figure out how you can manage to do what I've requested of you."

Add these new body language descriptions to the other three.

Also, you may recall other body language messages of your own. For example, John might say, "Mary, it occurred to me that when I like what you're doing. I wink at you."

These self-recalled behaviors also are added to the positive list.

Assignment 12

The behaviors you described in writing in the last assignment concerned your spouse's positive body language messages. Today you will begin to learn about your own negative body language messages.

THE ASSIGMENTS

1. Do The Second Exercise: Your Own negative Messages."

2. Study "Increasing Your Body Language Repertoire."

3. The "Always and Never" exercise.

4. Continue carrying out your Intimate Time ritual.

The Second Exercise: Your Own Negative Messages

Each spouse will think about, and try to visualize his/her own body language behaviors for, the five *negative* body language behaviors listed below:

1. "I do not believe what you're telling me."

2. "At this moment I do not like you."

3. "I disapprove of what you're doing now."

4. "You're being very unreasonable, and therefore I will not cooperate with you."

5. Any other negative behavior of your own which you can recall.

We want you not only to visualize the negative behaviors, but to be able to act them out at will. You are to act them out so realistically that your spouse will be able to identify them.

One way to learn to act out the negative behaviors is to know how they appear to your spouse, and then practice them. This can be done by watching yourself in a mirror.

Stand alone in front of a mirror (imagining you are talking to your partner). Stand far enough back so that you can see your hands, shoulders, and face. Then speak the behavior. For example, say aloud, "Mary [or John] I don't believe what you're telling me."

Say it loudly, vigorously, emphatically. At the same time, closely watch your body motions, gestures, and facial expressions. If not much happens, make up movements and expressions that express disbelief. Act out the body language message as best you can. Repeat this until the behavior appears to be clear even to a stranger witnessing it.

Do this for each of the five negative messages.

When both spouses individually have completed this, each will demonstrate to the other the negative body language messages as they have been rehearsed and self-observed.

The spouse observing the acting out may interrupt at any time to contribute impressions. For example, when John is demonstrating how he behaves when he does not like Mary, she might interrupt with, "Gee, John, I know you don't like me when you look at the ground when I'm talking, or turn away without replying."

Then she demonstrates the behavior, showing how she believes John usually displays this behavior.

Another example: When Mary is acting out one of her negative behaviors, John might interrupt, "Mary, I know you won't cooperate when I see you move your head sideways in little jerky motions and thrust out your lower lip."

Then John acts out the way he recalls having seen Mary do this.

Increasing Your Body Language Repertoire

After the five negative emotions have been acted out by each spouse, sit down together and have a body language "jam session." During the jam session each spouse contributes (as s/he thinks of them) any other non-verbal behaviors of any kind— either positive or negative.

Example 1: "I know you're grumpy in the morning when the corners of your mouth are turned down and you continually crack your knuckles."

Example 2: "I know when you're very happy in the morning because you hum and walk quickly, taking extra-long strides."

Example 3: "I know that if we're at a party and you want to go home, you begin rubbing your hands together and licking your lips."

Example 4: "I know that when you are skeptical about something you raise and lower your eyebrows and move your head slowly from side to side."

Each of you should list at least four behaviors similar to the ones given in the examples. They can be either negative or positive. Start with variations of our examples, if you want.

Discuss the body language messages just listed.

Demonstrate them.

By now you have accumulated a usable body language vocabulary. In the next assignment you will begin practicing it.

The "Always and Never" Exercise

The following small exercise concerns a negative message common in unhappy marriages. It is of special interest here because this particular negative message almost always is sent simultaneously by speech and body language; and both avenues of communication carry strong negative signals. The message involves using the words "always" and "never" when discussing your spouse's behavior or intentions.

An example:

When Mary and John were discussing their monthly budget, Mary said, "John, I've added twenty-five dollars for emergencies. I've got to do this because in the past you've *always* been so inflexible that we get into a jam month after month. You *never* understand that with two children. . . ."

In using "always" and "never" the speaker expresses negative expectations and lack of trust. The "always" and "never" statements define the other spouse as incurably wrong—a person incapable of doing well.

Because "always" and "never" exhibit negative expectations, the other spouse's behavior (as well as your own) will be directed into negative channels. *The way you expect an event to turn out directs both spouses' behaviors in that direction.*

In the example, Mary might have said, "John, we have a problem. I find it difficult to make our monthly budget balance because sometimes we have unexpected expenses which we can't avoid. This month, the school slugged us with an eighteen-dollar charge for the childrens' music lessons. They hadn't told us there'd be a charge.

"It would make my life easier if we could include twenty-five dollars in the budget for emergency items. If we don't need it we can put it into savings. Can we manage this somehow?"

Had Mary spoken in this way, her husband would have been more likely to agree her proposal.

The Exercise

Each spouse make up two statements about the other, using the words *"always" and "never"*. For example, "You're *always* late. You *never* consider that my time also is valuable."

Alternating, each says his/her "always" and "never" statements to the other. Emphasize the statements with gestures and facial expressions.

When this has been done, tell each other how each felt after hearing and seeing the "always" and "never" statements.

Then rephrase the statements, eliminating "always" and "never." Replace them with words that show respect and caring—bearing an overtone which says, "If we cooperate we can solve all problems." Emphasize this positive approach with appropriate gestures and expressions—body language signals which indicate that you cherish your spouse.

Example:

"You're a very busy person with enormous responsibilities, and I understand that you can't avoid being behind schedule sometimes. However, it would help me if we could work out some way for us to do things together more on schedule. A lot of it is my responsibility. *What can I do to make this happen?"*

When the statements have been rephrased in a positive manner, tell each other how each felt with the new approach. Contrast the feelings with those experienced after the "always" and "never" statements.

Assignment 13

Kinesics must be more than just exercises practiced while carrying out this program. Like all other exercises in the book, the ones concerning body language are designed to be used the rest of your lives. Assignment No. 13 is the first step toward day-to-day control of your non-spoken communication.

THE ASSIGNMENTS

1. Use body language to "talk" to each other (remember, the children are included).

2. Continue carrying out your Intimate Time ritual.

Using Body Language to Talk with Each Other: The First Big Step Toward Total Communication

From the last two assignments, you both have acquired lists of *how* you transmit certain behaviors and emotions to each other via non-verbal language.

Today you will perform these kinesics for each other, one at a time. However, you will not announce what your pantomime means. It is up to the observer to identify it—the way one does in playing charades. We suggest you start off with simple, uncomplicated behaviors.

For example, assume that Mary is the one to start off. She stands in front of John. Without saying anything, she wrinkles her nose and smiles.

John then tells Mary what he believes her body language statement means. He says, "You're telling me you love me."

Mary says that he is correct. Now it is John's turn to perform.

He stands in front of Mary. He cocks his head to the right, smiles, and looks directly at Mary.

Mary says, "That means you love me."

Continue this procedure until both of you can go through your entire repertoire (all of the behaviors and emotions from your lists, both positive and negative). Sometimes the one who is "watching" will not understand what the sent message means. In that case the "performer" must repeat the body language statement as often as needed, until the "watcher" correctly identifies it.

That's today's assignment.

Within a week, if you practice every day, you will have a repertiore of about twenty body language behaviors which are precise—that is, the "sender" of the message knows exactly what s/he is sending non-verbally, and the "receiver" understands what the "sender" is telling him/her.

You then will be able to use your communications "shorthand" even among other people. You will be able to exchange information which most other people do not understand. You will have developed a new aspect of intimacy.

What you are acquiring during the three body language assignments is but a small percentage of the vast kinesic vocabulary available to you for *willful use.* * We remind you again that within the marriage there is more information exchanged through body language than through words, and that even if you send information verbally, it usually is refined by the accompanying kinesics. For example, suppose you answer a person's question with, "I don't know."

If you say, "I don't know," with an accompanying smirk, the message may mean, "I really know, but I don't want to tell you".

If you scratch your head and thrust out your lower lip, it might mean, "I really don't know, but it is an interesting question, and I'd certainly like to know the answer myself."

If you run your finger across your nose and turn your head away, it might mean, "I consider your question improper."

Experience shows that there is no need for us to urge you to practice your body language. It is a new toy which is fun to play with. Of course, it is much more than a toy.

---

*For those interested in further information on this subject, we suggest the following books: Ray L. Birdwhistell, *Kinesics and Context (Essays on Body Motion Communication)* (Philadelphia: University of Pennsylvania Press, 1970); Albert E. Scheflen, M.D., with Alice Scheflin, *Body Language and the Social Order* Englewood Cliffs, N.J.: Prentice-Hall, 1972).

Assignment 14

You (like everyone else) have many behavioral habits which you perform mechanically without being aware of them. There may be some which you originally intended as positive messages but which your spouse interprets as negative. Likewise there are messages which may appear negative to you but which your spouse intends as positive. There may even be messages you intend as negative but which are received as positive.

The Red and White Bean exercise is designed to (1) make you aware of your unconscious behaviors, and (2) to assist you in reducing the behaviors your spouse interprets as negative and increasing the ones s/he interprets as positive.

THE ASSIGNMENTS

1. Do the Red and White Bean exercise.

2. Continue your Intimate Time ritual.

The Red and White Bean Exercise

Get approximately twenty-five red beans and twenty-five white beans, or twenty-five red poker chips and twenty-five white poker chips; or any small objects of two different colors.

Put the beans into a box. Next to this place two empty jars, one marked "Red Beans" and one marked "White Beans."

The Exercise

It will require about an hour to learn how to do this exercise. Once you know how to do it, continue practicing it for the rest of the day. It will not interfere with your normal routine, no matter where you are or what you are doing.

How to Do It

When your spouse behaves to you in a way that you experience as pleasant or positive, inform your spouse that s/he has pleased you.

The way to inform him/her that s/he has pleased you is by giving him/her a white bean.

The pleasing behavior can be a smile, a supportive statement, a favor rendered, or any behavior which suggests, "I care for you" or "I respect you."

When you give a white bean to your spouse, you also must thank him/her for the specific behavior. For example:

"Mary, thanks for the wonderful game of tennis."
"John, thanks for the flowers."
"Mary, that look you just gave me makes me feel wonderful. Thanks."

"John, thank you for cleaning up the garage."

On the other hand, when you experience an unpleasant or negative behavior from your spouse you also inform him/her of the fact—this time the red bean is given.

An unpleasant or negative behavior can be an unfriendly criticism, a smirk, a sarcasm, an angry interruption, a promise not kept—any behavior which you perceive as meaning "I disapprove of you," "I don't like you," "I'm trying to annoy you," or "I don't respect you."

The person who received the red bean *must* ask why. Don't assume you know; you may be wrong.

The person who gave the red bean *must* tell, courteously and calmly, why s/he gave the red bean. The receiver listens and acknowledges but does not debate the issue. That's all there is to it. Whether the receiver believes s/he deserved a red bean or not is irrelevent. *The only significant thing is the information that s/he has done something which the other spouse interprets as negative.* There are no further discussions. There is no arguing, no excuse-making.

Examples:

Mary hands John a red bean. He asks, "Why the red bean?"
Mary replies: "I said my father is a smart man. You pursed your lips and raised your eyebrows. When you do that I interpret it as meaning you think I'm full of hot air or lying."
"OK, Mary. Thank you for telling me."

John hands Mary a red bean. She asks, "Why the red bean?"
John replies: "The children asked me to play ball with them. You answered, 'Daddy's too busy now.' I appreciate your looking after my interests, but if a question is asked me, I prefer to answer it myself."
"Thanks for telling me, John."

The above examples concerned negative behaviors. In each instance, the receiver of the bean asked why s/he received the bean and the giver of the bean gave the reason.

However, in the instance of positive behavior, the giver-of-the-white bean immediately says why s/he gave it, and also gives thanks.

Mary hands John a white bean. She says: "John, I loved the way you smiled at me. Thank you."

John hands Mary a white bean. He says: "Mary, I came home tired and dejected. You said, 'Let's sit down and talk about it.' I said I didn't have time, we had to rush off to the Smiths' party. You kissed me and said, 'You're more important than the Smiths.' Thank you, Mary, you really made me feel good."

Mary hands John a red bean. Upon being asked why, she explains: "You said: '*Why* must you *always* smoke so much?' I appreciate your concern over my health; however, starting a personal questions with *why*, and saying *always*, puts me on the defensive."

After receiving either a red bean or a white bean from your spouse, put it in the appropriate jar. One jar is for the red and the other for the white.

Note: Be generous in handing out the white beans. It is an effective way of letting your spouse know that you are aware of and appreciate his/her positive and pleasing behaviors.

Scoring the Red And White Bean Exercise

At the end of the day, count the total number of white beans and the total number of red beans which have been given. *No individual scores are kept.* Your only interest is the total number of positive behaviors and of negative behaviors which have been performed by both spouses that day.

It is only the total number of each color which is significant. The totals are put on the scoreboard daily (see Figure 1). Tape the scoreboard to the front of the refrigerator, or put it in any other prominent place where everyone in the family can see it.

From day to day connect the marks on the scoreboard, thus

FIGURE 1

Mark number of white beans for each day with "o": o—o—o
Mark number of black beans for each day with "x": x—x—x

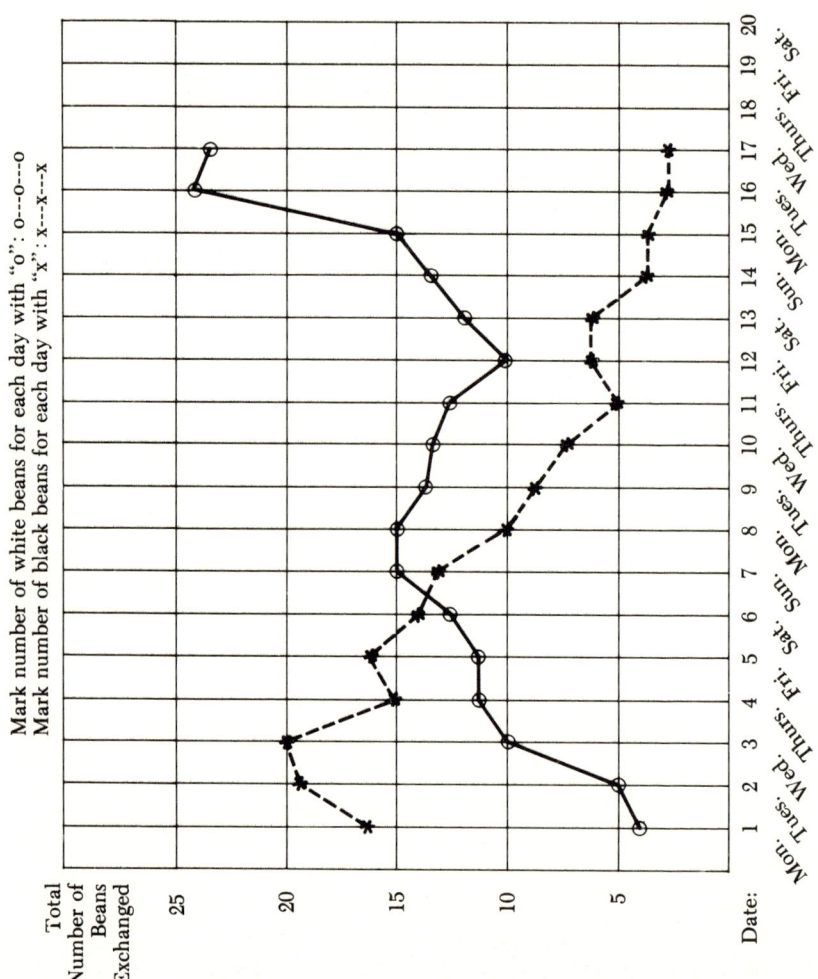

making two graphs, one for the red beans and one for the white beans.

Note: The graphs on Figure 1 show the average scores of twenty-one couples who took our course. Most couples start off with considerably more red beans (behaviors experienced as negative) than white beans (behaviors experienced as positive).

On the average, the turning point (when the white beans outnumbered the red) was the seventh day. Usually the positive behaviors (white beans) are in the preponderance from this point on.

(When you go out together, put some beans in your pockets or purse. If your spouse does something which pleases or annoys you, quietly, privately, and without fuss hand her/him the appropriate bean.

Do not delay the bean-giving until after you are home. [Exchanging beans when the action is fresh in both minds informs the other in a quiet, dignified way. And if a red bean is given, the receiver has a short reflective period which frequently eliminates the urge for sharp words.)

The Children

From the first day, the children will be curious about what is happening. Bring them into the exercise on the second day. Children from about the age of five upwards are capable of participation. The beans should be exchanged among all members of the family between parents, between children, and between parents and children.

The first few days after the children start, they may give red beans with a heavy hand. This must be handled with understanding and patience. It usually takes about four days for this heavy handedness to level off.

We suggest that when the family's total number of white beans exceeds the total number of red a celebration be held. Decide by democratic process whether it will be a movie, a family skiing day, a picnic—or whatever the majority decides.

Continue the exercise for at least ten days. The process may become part of your way of life. When this occurs, it is not uncommon to hear expressions such as, "This has been a white bean day!"

AN ASIDE BETWEEN
ASSIGNMENTS

There is a story of how in Denver, several years ago, a caustic chemical accidently got into a tank-car of gasoline. Before it was discovered, a number of automobiles had been filled with the polluted gasoline.

One of these cars was owned by an automobile mechanic. When his car began acting strangely (there was a loss of power and occasional spluttering) he took the car to his own garage. He examined the spark plugs, the carburetor, and the distributor. They all seemed clean and functioning. He tested the car again. It still spluttered and was low in power. The mechanic consulted with a colleague who also worked the car over, tracing out the electrical system and cleaning out all the pipes and ducts through which the gasoline flowed.

It was not until they drained the gasoline and put in a fresh tankful that they discovered the source of the trouble.

Spluttering marriages are analogous to the car in the story.

There comes a time when the unhappy partners both admit that their relationship is spluttering and has lost its power. However, they are unable to figure out what is messing up their marriage. They feel helpless—and begin to blame each other. Yet they still want to stay married.

Some will consult psychiatrists, psychologists, the clergy, or marriage counselors, or talk the ears off close friends. Some— in their anguish or perhaps in the hope of a quick miracle—go to encounter groups, ashrams, primal scream groups, gurus, and a variety of cults.

Despite these frantic activities, *the majority of these un-happy marriages continue in a state of habitual discord, or they terminate in divorce.*

An extensive survey conducted in 1963 by Don D. Jackson, M.D., Director of the Mental Research Institute in Palo Alto, indicated that:

1. Only 11 percent of the couples who had marital counseling (over a long period) reported that their marriage had improved permanently as a result of their therapy.

2. 40 percent of the couples reported that marital therapy had effected little or no change in the quality of their relationship.

3. The remaining 49 percent were divorced.

Certainly the general failure is not a matter of intent. Most couples who go to therapists believe that they want to improve their relationship, and the therapists sincerely want to help the couples. Yet the failures occur even though the therapist is well trained and does a good job of cleaning the emotional spark-plugs, the communications distributor, and the psychological carburetor.

In the lives of the partners there may be hidden destructive elements or influences—behavioral pollutants—of which nei-ther the therapist nor the couple is aware. These elements may include subtle physical ailments.

In Part II you will learn methods of identifying and eliminat-ing the destructive elements common to some unhappy rela-tionships—the physical problems of which some couples may be unaware. These subtle ailments can make any psychological therapy more difficult.

Assignment 15

*For spouses to have total communication (understanding),
each must be aware of the other's inner feelings. Having
this knowledge on a day-to-day basis is a subtle but very
powerful aspect of a satisfying relationship.*

*The exercise in communicating one's feelings is designed
to promote the exchange of "how I feel" communications.
Children should participate in this assignment.*

THE ASSIGNMENTS

1. Do the "Communicating One's Feelings exercise.

2. Continue the Red and White Bean exercise.

3. Carry out your Intimate Time ritual.

The "Communicating One's Feelings" Exercise

The spouses, equipped with a magazine or art book, flip a coin to see who starts. The winning spouse chooses a picture and looks at it for a little while. Then, putting the picture down, s/he describes it from memory to the other spouse (and to the children, if there are any). The description should be as detailed as possible.

After describing the picture, the spouse tells what feelings were stimulated by looking at it.

Example:

Mary wins the toss. She chooses a reproduction of a medieval painting of a mother and baby. Mary describes it as follows: "It is a picture of a mother and her baby. The mother has a blue dress with white lace on the collar. The dress is loose at the left shoulder. The naked baby is taking milk from the mother's breast. The half-smile on the mother's face says, 'I love you and will take care of you.' While taking the mother's milk, the baby's eyes are turned up to the mother's face. The eyes seem to say: 'I trust you. You will look after me.' The picture gives me a feeling of warmth and goodness. I feel as if perhaps I am the woman in the picture and the baby is our baby. I wonder, 'Why didn't the artist put the husband in the picture also?' "

When Mary has described the picture and her feelings, John repeats what he heard Mary say about the picture and her feelings—*exactly as he recalls what Mary has said.*

Mary listens to John, and if he has not recalled everything she said earlier, or has made mistakes, she corrects him in a kind, uncritical way. She also tells about any further thoughts she may have had about the picture.

Now it is John's turn. He choses a different picture. He describes it and his feelings about it. When he is through, Mary repeats back what he has said, and then, if necessary John corrects her.

If there are children, they go through the same process—with one of the parents or a brother or sister doing the listening and repeating.

For most people this will be a new experience, and, therefore it should be repeated several days in a row. After the first day, the object need not be a picture. It can be anything—a statue, a street scene, a flower, a view.

This exercise is designed to give couples an opportunity to practice subtle levels of communication with neutral subjects. Couples who have conflict over particular communications involving their feelings usually are willing to risk self-expression in this non-threatening way. The exercise will work effectively only when the repeat-back is done. At the start there will be some incomplete or incorrect recollections of what was said, but the spouses will realize that these are *not* signs of lack of interest. Rather, they are an indication of the need to develop better means of expression and better listening. It will hearten spouses to experience the improvement in their ability to both express a reaction and to listen to another's that will come with repeating this exercise.

Assignment 16

Very soon—starting with assignment No. 18—you will begin an exercise which can contribute much toward "liking each other." This is the Performance Pacts exercise. In this exercise you will be making many requests of each other and asking each other many questions.

Today's assignment is designed to assist you in making requests which are likely to be granted, and also to assist you in asking questions which will give you the information you desire.

THE ASSIGNMENTS

1. Study "The Art of Asking Questions."

2. Study "Rules for Making and Granting Requests."

3. Continue exchanging red and white beans.

4. Continue your Intimate Time ritual

The Art of Asking Questions

In a marriage, questions and answers provide more than an exchange of direct information. Questions and their answers powerfully influence the quality of the relationship. The substance of the questions, and the way they are asked, can bring about cordial cooperation or abrasive hostility. Further, the questioning process clearly indicates whether either spouse is attempting to dominate the relationship.

The question "When will you do as I ask?" indicates that the speaker may be trying to establish that s/he is in charge.

However, if the question had been "Would it be convenient for you to help me move the furniture next Friday evening?" it would have reflected an attitude of respect and equality.

There are several rules for question-asking and answering which (1) provide spouses with a maximum exchange of information, and (2) improve the quality of the relationship.

RULE 1
Phrase questions specifically and simply (not in complex generalities) so that the questions can be answered in the same way.

A general question: "What do you think of the political situation?"

A specific question: "Whom do you want to be our next president?"

A general question: "What do you think about Mike doing chores and getting an allowance?"

A specific question: "I believe Mike should do an hour of chores every day and get a five-dollar-a-week allowance. What do you think?"

A Short Exercise

Each spouse make up and ask two general, complex questions. The other spouse answer them. Then, rephrase the questions so that they are specific and simple. The other spouse answer the rephrased questions.

RULE 2
Avoid asking questions the intention of which is to manipulate the listener instead of getting straightforward information.

Example: John wants to change the arrangement between himself and Mary regarding the cleaning and cooking at home. John might manipulatingly ask, "Mary, at what time do you have to be at school in the morning?" or, "Mary, why do you have to be at school so early?"

The question is preparatory to asking her to take on more chores in the morning.

Mary might answer naïvely the first time or two and find herself manipulated into taking over an increasing share of John's agreed-upon responsibilities. If John frequently used this devious approach, Mary would withdraw more and more from his manipulatory conversation.

John should be straight and direct: "Mary, I find it hard to vacuum the house in the morning. Would it be possible for you to do this? Then I wouldn't have to rush so much to get to work on time. Perhaps I can take over one of your chores at night."

A Short Exercise

Each spouse make up one manipulating question and ask it of the other. Then rephrase it so that it is direct and straightforward.

RULE 3
When asking questions about your spouse's opinions, desires, beliefs, or feelings, make certain you accept the responsibility and ownership for your questions. This can be accomplished by *stating your own position before asking your spouse to answer the question.*

Wrong question: "What do you think about increasing Tom's allowance?"

Right question: "I notice that Tom has taken on a lot of extra chores. I believe we should increase his weekly allowance. What do you think?"

A Short Exercise

Each spouse should make up one question concerning the other's opinion, desire, belief, or feeling. Do not state your own position.

Then rephrase the question so that your own position is included.

Rules for Making and Granting Requests

Making requests of each other is a form of questioning. There are additional rules for asking and granting requests.

The quality of a relationship is clearly indicated by the number of things which are requested of each other by the spouses, the nature of the requests, and the way the requests are made and responded to.

Requests in a good marriage are reciprocal commitments to the relationship. Both spouses, by asking and granting requests, are putting deposits into a jointly owned bank of good will.

The closeness of the relationship is indicated by the number of positive requests for small but pleasing behaviors and services. When relationships deteriorate, the first requests to go are the ones keyed to pleasurable behaviors. Disengaging wives and husbands seldom ask each other, "Please rub my back," or, "Would you bring me my coffee in bed this morning?"

If the breach in a relationship continues to increase, the negative requests gradually outnumber the positive ones. The negative requests usually concern complaints and attacks: "Will you please stop criticizing me all the time?" or, "Will you please stop nagging me and give me a little peace?"

Even if an outside observer does not know the content of the request, s/he can tell whether a couple is emotionally separating just by observing the manner, the postures, the gestures which accompany requests. Fists are clenched, faces grimace, speech is angry. Over time, "Please" degenerates into "Will you?" and then to "You should," "You must!" and finally "Why the hell haven't you?"

When bottom is touched, all requests stop and spouses pass each other in wordless anger.

The kinds of requests the spouses make to each other and the way in which the requests are made, therefore, indicate the attitude of the spouses to themselves, to each other, and to their relationship.

RULE 1
If you have a request, make it. Don't remain silent, expecting your spouse to be a mind-reader.

RULE 2
A request is vastly different from a demand. If you want something from your mate, ask for it as a favor regardless of what it is. A request means "Please help me," and "I value you." A demand implies that the speaker is superior and is in charge.

RULE 3
The request must be reasonable, possible, and timely.

Spouses often expect (and request) more from their mates than is reasonable, or perhaps possible.

Example: While Mary is making her daughter's breakfast and rushing about to get to her office on time, John suddenly wants to take a sandwich that day instead of eating out. He asks Mary to make a sandwich for him: "Mary, I want to work through lunch today, so I'll take a sandwich. They taste better when you make them; would you do that for me, please?"

John's timing is wrong, and he's made an unreasonable request—even if the form of his request is perfect. The wise action would have been to make his own sandwich, or to relieve Mary of some of her chores so that she would have the time to make the sandwich.

RULE 4
Make certain that the request is not a mini-double-bind request.

A mini-double-bind request is a request phrased in a double-negative manner and thus conveying a confusing message. A few examples:

"You don't want to go to the store for me, do you?" (The intended message is, "Will you go to the store, please?")

"You don't want to take me to church, do you?" (The intended message is: "I'd like you to take me to church. Will you?")

In shaky relationships, the listener is inclined to answer the above double negatives with, "No, I don't."

Double-bind requests indicate that the requestor is behaving *as if* s/he expects a negative answer.

It is more effective to be direct, positive, and to accept the responsibility of what you are requesting. For example, "John, I'd like to go to church. I'd like it if you'd come with me, will you?" John can say "No", but he knows clearly what Mary would like and has requested. She has expressed her own very unambivalent positive desire.

The rules given above concern *making* requests. There are additional rules for *granting* requests.

Rules for Granting Requests

RULE 1
Agree to only those requests you can carry out. *Have the courage to say no to a request when no is appropriate.*

Sometimes spouses say yes to requests which they know they might not or cannot carry out. Example:

John has to rush frantically when he leaves work at five. He has to pick up the other four members of his car pool, then pick up Julie at dancing school, and then get home to help Mary get things ready for their dinner party, which starts at seven.

When Mary requested—despite his frenzied schedule—that he pick up some flowers for the table, John might have said yes, wanting to please Mary—even though he didn't see how he could possibly get to the florist shop before it closes.

That night, when Mary is setting the table for company and she has no centerpiece, she is justified in being annoyed at John. *She had requested that he pick up the flowers, and he had said that he would.* He had agreed, even though he knew he might not be able to carry out the request.

RULE 2 (For Request Granting)
Do not accept all requests as yes-or-no requests.
Treat a request as a negotiable action.

Requests should not always be considered as absolute and complete. Some can be fulfilled only in part, or perhaps in ways different than asked for. Doubtful requests should be discussed openly; and in an open manner be accepted, negotiated, or rejected.

When John asked Mary to make his sandwich, Mary might have said: "No. I haven't time. Please make it yourself."

Or Mary might have negotiated: "I haven't got time. However, I'll be glad to make it, if you'll finish cooking breakfast for Julie, and then go out and start my car so it can warm up."

Or John, as the one making the request, could have perceived that the request was unreasonable and untimely—and started the negotiating when he made the request: "Mary, your sandwiches taste better than mine. If you'll make me some, I'll take over what you're doing."

Assignment 17

Earlier, in Assignment 3, you learned that behavior is everything you do, and that there are three broad levels of behavior. In Assignments 5 and 6 you worked with the first behavior level—the molecular behaviors (the Cherishing Days exercise).

It is now appropriate to move up to the next level—that of performance behaviors. In this assignment you will learn, in detail, what performance behaviors are, and you will then make preparations for requesting desirable performance behaviors from each other.

THE ASSIGNMENTS

1. Study "Performance Behaviors: A Definition."

2. Make preparations for carrying out the performance pacts.

3. Repeat the exercise in communicating one's feelings.

4. Continue exchanging red and white beans.

5. Continue the Intimate Time ritual.

The Performance Behaviors: A Definition

The molecular behaviors are the small transitory behaviors—a quick smile, a squeezing of the hand, a frown, the bringing of a cup of coffee in bed, a turning away, a wink.

The next level consists of the performance behaviors. These are the middle-sized behaviors, those used by spouses in:

1. Assisting each other in doing the daily unavoidable chores.

2. Helping each other enjoy recreational activities.

3. Collaborating on the sometimes tedious responsibilities of civilized living.

4. Redesigning his/her personal tastes to please the other spouse while at the same time maintaining his/her uniqueness.

Whereas the molecular behaviors establish the general tone of the relationship, the performance behaviors largely establish the way the spouses think and feel about each other. Naturally the spouses desire a continuum of behaviors which stimulate both of them to think and feel towards each other in joyful, satisfying, and positive ways.

The connection between this goal and the performance behaviors can best be explained by examples:

Mary believes that she would feel better about the marriage if John would recognize and encourage her desire to go to night school so that she could get her master's degree and thus get a better job. His encouragement of her would be a performance behavior. If he were to look after the children during the afternoons and evenings that Mary goes to school, this would be a performance behavior.

Mary thinks that she would be more satisfied with their recreational time if John would sometimes take her to the opera, instead of his suggesting that she go with a neighbor. This would be a performance behavior.

She believes that she would be more respectful of John if he helped more with the household and family chores, and if he

made her parents feel more welcome when they came for over-
night visits. These would be performance behaviors.

Mary is concerned with the way John "performs" in the relationship.

John believes that he would feel better about the marriage if
Mary expressed interest in his business and offered to help plan
his sales campaigns or perhaps do some of his book-keeping.
These too would be performance behaviors.

He thinks that he would feel more satisfied with the time
Mary and he are together if she took the initiative in gaining
skill in some of the things he enjoys—for example, singing,
playing bridge, and hiking. These are performance behaviors.
He also believes that he would think more of Mary if she
learned to be more independent when she makes decisions
about choosing things like curtains in the kitchen or clothes for
herself, or an after-school program for the children. These
would be performance behaviors.

John is concerned with the way Mary "performs" in the relationship.

Mary would feel more comfortable if John would attend to
her more when they are at parties or public gatherings together. This would be a performance behavior.

John would feel more comfortable in the marriage if dinner
were a leisurely affair, with Mary as well groomed as she was in
their courtship days. These are performance behaviors.

Performance behaviors, then, concern *how each performs
in the routine of the relationship.* The performance behaviors
are present from getting up in the morning to feed the baby
until late at night when the cat is put out and the doors are
locked.

There are many things to be performed in a relationship, and
they tumble forth in helter-skelter clusters. Therefore, in the
performance behaviors it is advantageous for the spouses to
agree on orderly rules for carrying them out. This is especially
necessary when the spouses come from different social, economic, or ethnic backgrounds, in which the daily living routines
and tastes may be different.

When there are orderly and agreed-upon rules, then Mary can depend on John's cleaning up the kitchen when she leaves early in the morning to drive the children to school; and John knows that Mary will set aside Friday night to help him type his reports.

The rules (which will be explained below) not only provide a sense of unity but also give the spouses the security of behavioral freedom. Agreeing on a set of rules for performance behaviors lets spouses know what performances to expect from each other, and how each will respond to the other's daily routine actions. Thus, by mutual agreement, both know when s/he is free to work late and miss dinner at home, or be tardy; and both know when and with whom each can have a few drinks outside the home.

Rules applying to performance behaviors can permit spouses to feel comfortable when they are apart, and free of strife or indecision when they are together.

All these satisfactions can be realized when spouses:

1. Inform each other what frequently done performance behaviors each wants from the other, and how these behaviors are to be performed.

2. Negotiate the above, so that both spouses get most of what they want from each other and can perform in the marriage as equals.

These conditions can be accomplished by making a Performance Pact.

Performance Pacts: Step One of Preparation

In the first step of preparation for the Performance Pacts, you will begin thinking about the performance behaviors you would like your spouse to do.

The performance behaviors which you want your spouse to do for you, and which you will request of him/her must meet six precise requirements:

1. The request must be specific, not general.

2. The request must be a service or behavior which can be performed at definite and specified times.

3. The request must be for a behavior which can occur frequently.

4. The request for a service or behavior must tend to improve the quality of the relationship for both spouses.

5. The request for a service or behavior must be of such a nature that the person who does the requesting can assist the other in carrying out the desired action.

6. The request must be stated in a positive, not a negative, manner, as a courteous request, not a demand.

Example:

> Would you go with me to the weekly plays in Burlington? There's a different play on every Tuesday night. The series lasts two months. We will have to leave home at about half-past six and will be back around midnight.
>
> I know you're often rushed in the late afternoon, so I'll be glad to prepare dinner on Tuesday evenings and arrange for the baby-sitter. This will give you time to bathe and dress. I believe we can have a lot of fun together at plays.

Note that:

1. The request is specific—to go to the plays together.

2. The behavior is to be performed at a specific time—Tuesday nights between half-past six and midnight.

3. The behavior is something which can be done frequently —once a week for eight weeks.

4. The behavior will tend to improve the relationship for both spouses, since they will both have a good time.

5. The spouse making the request can and will assist the other spouse in carrying out the request.

6. The request is positive and is phrased courteously.

More examples:

> On Saturday morning, after the boys have gone to their scouting meeting—about 9:00 A.M.—would you help clean the house? You could be responsible for all the upstairs rooms. It will probably take you about three hours, which is how long it usually takes me.

In order to help you, I will have the boys change their sheets and make their beds before they go. I'll persuade them to be neat and not drop things all over, and I'll make certain that all the cleaning equipment is assembled at the top of the stairs.

Would you be willing to attend the one-semester nutrition course being given every Monday night at the high school from seven until ten o'clock and then teach all of us what you learn?

I believe that if we had the kinds of meals they discuss in the course, our family would become healthier. Also, our food bills might be lower.

There is a food store about three blocks from where Julie has her late-afternoon dancing class. When I pick her up, I'll be glad to do the shopping for things we will need. Also, if you will do this, I will stay home with the children on Monday nights, when you go to school.

Would you be willing to wear your hair shorter? It has a kooky look and I feel uncomfortable about it. When we got married I enjoyed your scrubbed, "all-American" appearance. You used to have your hair trimmed every two weeks. You say barbers are expensive, but if you'll go back to your old routine I will be glad to pay for the barber from my personal spending money, and I believe I'll be happier when we go out together.

Before you do anything else, sit down together and briefly *mention* some of the performance behaviors which you believe you may ask of your spouse.

At this stage do not check to see whether the six requirements have been fulfilled. This is just a preliminary exchange of ideas; and its purpose is to stimulate the flow of thoughts on desirable performance behaviors.

It is not a discussion. One of you simply mentions something you believe you may ask the other to do for you in the Performance Pact. The other listens, acknowledges, but does not reply. For example, Mary might say, "John, I wish you'd drive me to my office every morning so that I don't have to take the bus."

In this case John acknowledges that he has heard but does not say yes or no. He receives Mary's statement merely as a suggestion, a vehicle for stimulating his own ideas.

Exchange ideas in this manner for ten or fifteen minutes.

Performance Pacts: Step Two of Preparation

With the six qualifications in mind, *write* out the performance behaviors you want your spouse to do for you.

Write each request on a separate piece of paper.

The more requests you make of each other the better. There is no maximum number, but we suggest a minimum of five. However, the requests made by each spouse should be equal in weight to those of the other. (You will learn shortly how to measure the weight of the requests.)

When you have written out your requests, carefully go through them to make certain they meet the six requirements.

Then exchange papers.

Read the requests made by your spouse and think about them tonight and until you meet for your next assignment tomorrow.

Assignment 18

More preparation is needed before the Performance Pacts can get underway. In this assignment, each spouse will make certain that the other's requests fulfill the six requirements and that the requests are completely understood.

THE ASSIGNMENTS

1. Do Step 3 of preparation for Performance Pacts.

2. Continue exchanging red and white beans.

3. Carry out your Intimate Time ritual.

Performance Pacts:
Step Three of Preparation

Alternately, each read aloud your spouse's written requests for performance behaviors. After you have read one of your spouse's requests, analyze it:

Does it fulfill all six qualifications?
Do you completely understand it?
Tell your spouse how you understand her/his request.

What follows is the first draft of a request:

> I request that once a week you help clean the house. This would help me very much. Also, I believe it would give us a sense of doing things around the house together.

The request was read aloud.

In the opinion of the spouse whose services have been requested, not all of the six requirements have been met. Therefore, the following questions were asked:

"On what day would you like me to help clean the house?"
"During what hours, and for how long?"
"Is there any particular part of the house you want me to clean?"
"Also, what action can you take to make it easier for me to carry out your request?"

In this manner, both spouses go through all of each other's requests, reading them aloud alternately, and asking questions and changing the requests as appropriate to meet the six requirements.

What has happened so far is that the requests have been made, the six requirements have been complied with, and both spouses understand each other's requests.

However, as yet, neither has agreed to comply with each other's requests. It is first necessary to make certain that the exchanged behavioral gifts are approximately equal in value. Therefore, the requests must be "weighed"—given value ratings.

Assignment 19

Both spouses have—in writing—made requests of each other. It is necessary that the total value of the spouses' requests be approximately equal: the time and effort required to carry out the wife's requests should be approximately equal to the time and effort required to carry out the husband's requests.

In this assignment you will give a weight—a value—to each request. Each request must be rated according to the ease or difficulty involved in carrying it out, its convenience or inconvenience, its comfort or discomfort.

THE ASSIGNMENTS

1. Give a value to each request.

2. Equalize the requests as appropriate.

3. Study "Giving Thanks."

4. Study "What To Do If You Believe Your Spouse Is Not Carrying Out a Request Properly."

5. Continue exchanging red and white beans.

6. Continue your Intimate Time ritual.

Performance Pacts:
Step Four (Equalizing Requests)

Study each request which has been made of you and then assign a weight to it. Use the following scale (examples will be given shortly).

A rating of 1 for very easy tasks or behaviors, those which can be done comfortably, easily, and conveniently.

A rating of 2 for relatively easy tasks or behaviors which can be done with only slight inconvenience and which cause some, but not very much, difficulty or discomfort.

A rating of 3 for tasks or behaviors which are fairly difficult, which require considerable extra time or trouble, or which can make the person doing them a bit uncomfortable.

A rating of 4 for tasks or behaviors which are very difficult, and which cause considerable inconvenience and discomfort.

A rating of X for tasks or behaviors which are unacceptable —that is, the request is turned down. (All other ratings, from 1 to 4, mean that the request is acceptable.)

Examples:
One spouse requested that the other wake up the children every school morning and make sure they get up. This is easy for him/her because s/he is up early anyway. It will be convenient, and it may even be enjoyable. Therefore, the spouse receiving the request gave it a rating of 1.

One spouse requested that dinner be served every night in the dining room instead of in the kitchen. The spouse making the request offered to serve the meal in the dining room on the nights s/he prepares dinner, and requested the other spouse do the same when it was her/his turn. This was considered relatively easy to do; the only inconvenience concerned changing her/his habit of laying work papers on the dining room table for later studying. Eating dinner in the dining room would not cause any discomfort—only a bit more work. Therefore, the spouse receiving the request gave it a rating of 2.

One spouse has requested the other to clean house for three hours every Saturday morning, beginning at 9:00 A.M. The spouse who received the request had an established routine of going over the household accounts every Saturday morning. To clean house at this time would have required doing the accounts on Sunday mornings instead of Saturday; which, in turn would mean getting up to attend church services at 8:00 A.M. instead of 11:00 A.M.

The spouse receiving the request felt that granting it would be good for the marriage. However, since it required extra time, caused the discomfort of creating a new schedule, and was somewhat inconvenient, it was rated as 3.

One spouse requested that the other attend a nutrition class every Monday night and then teach the family to practice what was learned in class.

The spouse receiving the request had planned to join a social club which met on Monday nights. Joining could have been delayed until the three-month nutrition class was over, although there might not be a membership opening then. Also, the spouse receiving the request was a meat-and-potatoes person who didn't enjoy eating green vegetables; s/he also likes bread, cakes, pies, ice cream, etc. S/he realized that going to the nutrition class threatened the family's established eating habits, and thought that they might even turn into "health nuts."

The spouse receiving the request felt that granting it would help the marriage; however, it would cause her/him considerable inconvenience, would be emotionally difficult to do, and might cause provoke discomfort. Therefore the request was rated 4.

One spouse requested that the other not wear his/her hair so long, and that it be cut frequently.

The spouse receiving the request liked having long hair because of its looks and because it was in style; almost everyone at her/his place of business also wore long hair. Furthermore, most of the couples' friends knew that the other spouse had been nagging for months to have the hair cut shorter. The spouse receiving the request was sensitive about others gossiping about how the other spouse had forced him/her to cut the hair.

The request was given an X rating: it was unacceptable and was therefore turned down.

Balancing the Weight-Value of Requests

When every request has been rated, each spouse then adds up the ratings of her/his requests. The sum is the weight-value for the total number of requests that each has asked of the other.

The sum of the weight-values should be approximately equal. If not, *then the person who has received requests of too little value should ask for more.* Thus, if Mary's requests to John add up to a total of 20 and John's requests to Mary add up to a total of 30, John should ask Mary to request more performance behaviors from him. *Balance is always achieved by making additional requests rather than by canceling existing ones.*

Summary

Up to now, the following has been accomplished in preparing for the Performance Pacts:

1. Requests have been conceived and written for performances, services, or behaviors which will improve the quality of the relationship for *both* partners.

2. The requests have been written in positive and specific terms, both as to what is required and as to the timing.

3. Each spouse has noted in the request how he/she can facilitate the other's compliance with the request.

4. The meaning of each request has been discussed, clarified, and, if need be, revised for complete clarity.

5. Each request has been given a value, and the total has been added up. Further requests have been added, if necessary, to

create an approximate balance between what the two spouses have requested from each other.

When all the steps, as summarized above, have been completed, the spouses—perhaps with a handshake—grant each other the agreed-upon requests. Or, to say it another way, you and your spouse have agreed to exchange desirable performance behaviors for the assigned time, until the end of the rest of the course (about three weeks).

At that time you will renegotiate your Performance Pact, which is a short-term agreement. If you find flaws in it, you can change it three weeks from now.

Therefore, commencing tomorrow, you two will begin carrying out the Performance Pact you have just agreed upon.

Put the written requests in your common notebook. When your spouse carries out one of your requests, write the date it has been done on the paper on which the request was written.

Giving Thanks

Whenever a requested service has been done, it should be acknowledged with thanks. This is accomplished by:

1. The spouse who received the performance gift immediately gives thanks. ("Thank you for taking care of the upstairs this morning.") The other spouse acknowledges the thanks.

The "thank you" and its acknowledgment need not be phrased in exactly these words, as long as the meaning is clear and cordial. Reaching out and touching each other at this time can provide added warmth and meaning.

In the event that one spouse forgets to give thanks it is the responsibility of the other spouse to remind him or her.

It may appear that there are a great many "thank yous" and "you're welcomes" spoken. They may even cause some laughter. That is good. It will make both of you aware of how much you are doing for each other.

What to Do If You Believe That Your Spouse Is Not Carrying Out a Request Properly

If one spouse believes that the other is not fulfilling a request properly, s/he *should not criticize.* Instead, bring the subject up during the Intimate Time period. For example, one spouse might say: "Your cleaning the upstairs on Saturday is a big help. However, washing the windows is part of the cleaning job, and you haven't done that."

The other might repy that s/he will wash the windows from now on, or that s/he does not consider that part of the job as described on the request sheet.

In the latter instance, the first spouse should accept this and say no more, and express her/his thanks that the request in general is being carried out effectively and cheerfully. However, when the Performance Pacts are revised at the end of the course, s/he can make window-washing part of her/his new request.

Tomorrow you will begin to carry out your Performance Pact. It should be made for a duration of several weeks, after which you will renegotiate it. If you want to change any part of the pact, it is then possible. The agreements you have made are short-term commitments—so we suggest that you carry them out with vigor and good cheer.

Assignment 20

Today, concentrate on carrying out your Performance Pact. It may be that only one or two of the requested services or behaviors are possible today, the others being scheduled for other dates. If necessary, juggle your schedule so that each spouse can do at least one of the requested performances for the other.

In addition to saying "thank you" for the one or two requested performances which are exchanged, thank each other for everything *one receives from the other this day. Be alert for everything your spouse does for you or your family. Thank your spouse for preparing a meal, for bringing in the paper, for bringing a cup of coffee, for holding the door open, for bringing in the bundles, for the scores of services or satisfying behaviors which spouses contribute to each other daily—even in the worst of marriages.*

THE ASSIGNMENTS

1. Carry out your Performance Pact.

2. Thank each other for every service or satisfying behavior received.

3. Exchange red and white beans.

4. Continue Intimate Time.

Assignment 21

This assignment emphasizes (and will help you experience) a variation of the Law of Positive Expectations: When you define and describe your spouse's behaviors in a positive way (the way you would like them to be), it is likely that s/he will behave in that manner.

It is this attitude which, almost by itself, can change a relationship from a state of frustration to one of satisfaction.

You will observe that this assignment may involve role-reversal.

Although the assignment concerns a dinner party, spouses can apply the principle to all activities in the relationship.

THE ASSIGNMENTS

1. Plan the dinner party.

2. Carry out your Performance Pacts.

3. Exchange red and white beans.

4. Continue Intimate Time.

The Dinner Party

Plan a dinner party for a few nights from this evening. (Of course, if that is an inconvenient time, delay it for a day or two).

The person who will run the dinner party, the person who will be in charge of the dinner party, the one who does the work will be the spouse who usually does *not* do this.

If it is the wife who usually does the cooking, the food shopping, the cleaning up, and so forth, then it will be the husband who will run the dinner party.

If it is the husband who normally does the cleaning, cooking, and food shopping, then the wife will be the one who runs the dinner party.

If the norm is for both wife and husband to share these activities equally, then flip a coin to see who will be in charge. (Note: if wife and husband normally share all household chores and other marriage-community activities, the chances are you don't need this book).

For the purpose of explaining this exercise, we will assume it is the wife who does most of the cooking, food shopping, and home management. Having assumed this, then, it will be the husband who will be in charge of your dinner party. He will plan the meal, do the shopping, see that the house is in order, look after the children—feed them early if necessary—arrange for the babysitter, take the babysitter home. In short, he will do everything.

The wife, in this case, then, will swap roles; and she will do what the husband normally does in a dinner party situation.

What about the guests?

For the purpose of this exercise, the guests will be selected and invited by the spouse who is not running the party (in this instance, the wife). She will invite anyone she wants—without discussion or agreement with the husband. She simply tells him who is coming. However, we suggest it be limited to four guests.

It will be the guests of the wife's choice who will be impressed or unimpressed by the quality of the dinner. However, the wife is not to tell the guests that the evening is an exercise in marital improvement or an experiment. Simply ask them to dinner. We suggest she invite them immediately so that the cook-and-meal-planner will know for whom he must cook.

At this stage the spouses will engage in a bit of fantasy. The two of you will put on a small dramatic production immediately after the wife announces who the guests will be.

The play-acting will be that the wife (the one who is *not* running the dinner) will begin defining and predicting every-thing concerning the Saturday night dinner party—*in the nega-tive. In the negative* she will define and predict all of her hus-band's efforts to run the party. She will find fault with everything—two days before the event takes place.

(She will, in her negative statements, of course, be violating the Law of Positive Expectations, the "As If" Principle, and just about everything this book stands for. However this is play-acting. It is being done with a purpose—as soon will be appar-ent.)

Here are some examples of how the wife in this case defines and predicts things in the negative:

"Well, John, now that you're running the meal for a change, you sure look scared. Well, I understand your fright. You'll probably foul up everything and embarrass me in front of my friends."

The husband can answer in any way he wishes. However, he must go through with the dinner regardless of how nasty this discussion becomes.

The wife continues along these lines,
"John, I sure hope this place doesn't look like a pig pen when my guests arrive."

The husband can respond any way he wants.

"John, I hope the meal isn't hours late like your cook-out at your mother's was."

"John, please don't put us in a hole financially by trying to show off with a lot of expensive foods and wines which we never have for ourselves."

"John, when the guests are here, just because they're my

friends, please don't withdraw and get sulky or insult them the way you did when I had my bridge club here."

"I just can't wait to watch you muck around in the kitchen after the guests leave. Then you'll know how I feel when you stagger off to bed and leave me with a big mess to clean up."

And so forth. This temporary play-acting is the means by which the wife will define her husband, the party, and the relationship in the *negative*—whether she feels that way or not.

In her negative play-acting, the wife is behaving as if the dinner party her husband is preparing will be a failure. She is defining her husband as incompetent and thoughtless.

We know that if the scene were real rather than play-acting, her negative behavior would create a negative response. Her husband would be apt to behave incompetently and the dinner would end in disaster. This is the Law of Negative Expectations which, if followed, is the great destroyer of marriages.

In this play-acting we want the wife to act out her negative role with sarcasm, with fear of disaster, and with a hardly con- ·cealed contempt. We want the husband to react in any way he desires.

This small drama of negativity should last five or six minutes. We repeat—this negative performance is to be done now, at least two days before the dinner party.

When the playlet is completed, shake hands and congratulate each other on being talented actors.

After the above has been completed, we want you to go through a similar play—however this time the spouse who is not running the party (in this example, the wife) will fantasize and define everything concerning the dinner party *in the positive.*

The wife not only defines the party, her husband, and every- thing else concerning Saturday night in the positive—she does even more than that: she also states what she will do to help her partner make her positive predictions and opinions become realities on the night of the party. Her comments might be along these lines.

"John, it always makes us feel good when the house is neat and clean, especially when we have guests. You're a master at that, and I can hardly wait to hear the praises of our guests. Oh, if it'll help, I'll take the children to my sister's for the afternoon and evening; and I'll pick up some flowers for the table on the way home.

"You know, even though you haven't done much cooking, everything you've ever put your mind to, you've done well."

"John, our guests were invited for six-thirty, with dinner at seven. I'll bet it'll be the first time in their lives they'll experience a party which really clicks on time. Between us, if you like, we can concoct a menu which will let you do all the cooking the night before."

"John, cleaning up after a party always is such a bore, and you're so naturally neat that you may worry about this. I'll celebrate my "night off" and your big triumph as cook and host by helping after the guests have gone."

You need not use the above examples. Just make up an appropriate set of your own comments.

Summary

Go through the two playlets. In the first playlet, the person who is *not* running the dinner party is negative and defines everything as a failure. In the second, again the person who is *not* running the dinner party, does precisely the opposite. She/he not only defines everything in the *positive* and predicts success, but also says what she/he will do to make these *positive* predictions become realities. S/he says what s/he will do to *help,* not take charge and run the show.

Assignment 22

Most couples go into this assignment with a bit of nervousness. The person who is doing the work may have doubts as to whether s/he can manage the dinner party successfully. The other one may fear being shown up or that if the party is a failure, and that she/he will be embarrassed.

Don't worry. Every time we have observed this exercise done, it has been successful—as long as the one not doing the work behaves "as if" everything will fall into place beautifully.

THE ASSIGNMENTS

1. Prepare for the dinner party.

2. Carry out your Performance Pact.

3. Exchange red and white beans.

4. Continue Intimate Time.

Preparing for the Dinner Party

Today (in our example) the husband will make preparations for the dinner party he will be running tomorrow night. This is a party to which his wife has invited four of *her* friends or business associates.

Today the husband probably will be at his place of business all day and won't be able to start preparing dinner until late in the afternoon. However, he can plan his menu and go shopping for food. He may even do some of the cooking this evening. It is imperative that the meal be simple and well within the couple's normal dinner-party budget.

The wife should not help him directly with the cooking or serving. However, she can help him tremendously by making positive predictions, by assuring him that she is looking forward to an evening of fun, and by doing the things she has promised (in this case, the wife said that she would take their small children to her sister's house for the evening and get the flowers for the table).

Assignment 23

Today is the day of the dinner party. One spouse, and only one, has the responsibility of doing the organizing and the work necessary for a successful evening.

The other spouse has the responsibility of continually defining the situation IN A POSITIVE MANNER and doing whatever is reasonable to make the positive statements into a realtiy—without taking on the burden of running the party.

Have a good time!

We guarantee that when the positive-definition process has been carried out, the party will be a smashing success —even if John drops the bowl of curry sauce or some of the guests arrive three-quarters of an hour late. It will be a smashing success because the spouses will have the feeling of intimacy and supportiveness which usually results from defining each other in the affirmative.

P.S.: After the guests have departed, and after the children have been picked up and are home and asleep, and the kitchen has been cleaned up, extend your Intimate Time so as to have a private party for you two alone—to celebrate your success in proving that things go well when they are defined in the positive. Perhaps you will feel like discussing the principle: THE WAY YOU DEFINE YOUR SPOUSE AND THE THINGS HE/SHE DOES (OR IS ABOUT TO DO) IS THE WAY HE/SHE WILL TEND TO DO THEM.

THE ASSIGNMENTS

1. Enjoy the dinner party.

2. Continue carrying out your Performance Pact.

3. Exchange red and white beans.

4. Continue the Intimate Time ritual.

Assignment 24

Almost all couples who have been successful, so far, in improving their relationship by using this book, fall into a small trap. They begin to get smug. Because the relationship has become more satisfying (although there's still a long way to go), the spouses begin to take each other for granted. They may not be aware that no matter how wonderful the marriage is, both spouses are constantly going through the Marital Evaluating Process. That is, consciously or unconsciously, both are wondering, "Am I sufficiently satisfied with my spouse?" This assignment explains the marital evaluating process.

THE ASSIGNMENTS

1. Study "The Marital Evaluating Process."

2. Carry out your Performance Pact.

3. Exchange red and white beans.

4. Continue Intimate Time.

The Marital Evaluating Process (Am I Sufficiently Satisfied with My Spouse?)

All spouses *at all times* are influenced by many forces which affect the relationship. Some forces hold the spouses together, while simultaneously other forces tend to fragment the marriage. When Mary feels loved and cared for by John, the "stay-together" forces are strong and active. But when an old beau supplies much-needed empathy during a period when John is grouchy and working late, and Mary begins to compare her husband with the old beau, then separating forces may begin working on her. The old beau describes some of the good times they had together; and if he hints how much better life could be with him, the fragmenting forces may become stronger. Of course, the ratio between the stay-together and fragment-the-marriage forces goes up and down, depending on the moods and circumstances of the moment. However, the conflicting forces are *always* present in some form.

Mary's stay-together forces include the enjoyment of John's sense of humor and his musical ability. She feels his muscular body is a thing of beauty, and she gets pleasure from the way he becomes excited in describing his dreams of the future. These among many things draw her close to him. She also thinks that she would never find a better father for their children and realizes how strongly her whole family dislikes divorce. Thus for Mary the stay-together forces depend upon John's appeal and the problems which she associates with separation from him. The stay-together forces are, therefore, partly positive and partly negative.

The fragment-the-marriage forces are also a mixture of negatives and positives. Mary feels that John is a little less sophisticated than she would like; she finds his loud manner offensive at times; and, while she hates to admit it, she sometimes doubts that he will be much of a financial success.

She considers what life alone would be like, that is, if she and

John split up. She imagines the satisfactions of being her own boss (limited if she has the children). She visualizes divorce as a chance to be independent for the first time in her life. She sometimes likes the thought of meeting a variety of men, something which was denied her by her early marriage.

Thus, the forces which tend to fragment the marriage consist of a combination of dislikes in the present life situation and of possible gratifications in other situations.

The stay-together and the fragment-the-marriage forces are *both* always present in some degree. Partners in good marriages take both positive and negative forces into conscious consideration. Spouses who suffer in troubled marriages often stumble along, completely and unhappily convinced that their present spouse is undesirable and the cause of all the trouble.

In strong marriages, spouses know that just as they constantly evaluate their own reactions to marriage, their partners are making similar evaluations at precisely the same time. Therefore, they concentrate on increasing the positive value of the relationship so that their mate finds more appeal in staying at home than appears to be available in any other relationship. This process of strengthening the stay-together forces requires constant effort, just as does nurturing children, or looking after a flower garden. *However, the effort is well rewarded when spouses overcome their tendency to take the other's affection for granted and, instead strive to maintain a favorable balance between the counterpoised forces which would strengthen or weaken their marriage.*

For example, Mary and John (just like the rest of us, whether we are aware of it or not) automatically evaluate everything which happens to them. Some of the experiences within their relationship they would like to see continued or perhaps even increased, and there are other experiences they would prefer to have eliminated or diminished. *Their preferences are not abstract; they are comparative value judgments.* That is, the events they are judging to be good experiences or bad, desirable or undesirable, must be compared to some standard; and these standards usually are current relationships with other people, or previous experiences with people and activities, or with fantasies about people and events.

Before Mary married John, she had a composite of expectations about how much happiness and satisfaction marriage could offer. Her expectations had been formed partially by watching her parents' anguish as they suffered through an unhappy marriage which they felt they could not leave; the relationships of other relatives and friends whom she knew more or less well; her vicarious experiences through books and movies; and her fantasies as she imagined herself married to men whom she had known as briefly as one date or as long as many months.

All of these factors contributed to her aggregate "standard of satisfactions," which became her absolute index of measuring her joy in social relationships, including her marriage to John. Against this standard she had evaluated Richard, the man whom she was seeing when she met John. She had found that Richard provided about 60 percent of the satisfactions she thought possible. In contrast, she found over time that John offered her more than 70 percent of these sought-for joys. (She didn't really assign percentages to her suiters. However, in practical application she might have done so; and for the purpose of illustration we will use numbers as a convenient way of summarizing her feelings.)

John was less than perfect, she felt, because he would never be affluent, he had less education than she, and he came from a family a bit lower in social standing than her own. As she grew to know him better, however, she found more and more satisfactions in her relationship with him. She appreciated his strong manner. She liked the way his calmness helped her take things more lightly. She liked his self-acceptance and his generosity with her. Gradually her evaluation of her experience with him crept up to over 80 percent of her desired level of satisfactions. Before John, she had never come nearly this close with any beau.

Ten years after their wedding, John became stressfully involved with his job, and Mary experienced the pressures of being a working mother. During this period, Mary's level of satisfaction level fell fifteen points from the wedding day high of 80, down to 65 percent. This still was a little above what she had experienced with other men.

At about this time, a new family moved next door. Two months later the wife died, leaving Reggie, the widower, and his son alone in the house. Mary felt sorry for them, she cooked them a few dinners and occasionally helped Reggie out with his household chores. Reggie gradually began asking for help in other areas, from choosing the material for new slipcovers to advice on his son's problems in school or how to remove coffee stains from a tablecloth. From time to time, Reggie took Mary to dinner when her husband worked late or was out of town.

Mary observed that Reggie could afford to take her to places which John, her husband, could not afford; Reggie had traveled a good deal; and, despite the fact that his wife's death had hurt and confused him, he remained cheerful.

Mary's inner evaluating machine could not help comparing Reggie with John. Whether Mary's comparison was realistic or not is unimportant. She imagined him to be more caring than John and less demanding; she appreciated the way he accepted her as an equal, and she liked his polished manner.

In the old days when Mary rated her experience with John at 80 percent and her expected satisfaction with other relationships in general as 60 percent, John was safe and secure in her affections. But when the ratio became 65 for him and a 85 for the man next door, the relationship between Mary and John was in trouble, at least temporarily.

In the contemporary marital relationship, which is influenced by forces both inside and outside the marriage, the partners will remain in the marriage and stay committed to it when the relative balance of rewards is favorable. However, when the balance of rewards becomes unfavorable, then conditions are ripe for disengagement or for shifting some of the commitment to other people or other activities.

At all times, both partners are subtly, unwittingly evaluating their marriage. In successful marriages the partners accept this reality. Knowing each other's needs and desires, they work hard to provide each other with a high level of predictable satisfaction. When there are negative behavioral exchanges, they acknowledge them and negotiate change for the better. (This is the reason for the three major negotiation exercises in this book: Cherishing Days, Performance Pacts, and Authority Distribution.)

When spouses work at providing each other with a high level of satisfaction, the level of trust and satisfaction usually is higher within the relationship than it with other people or other activities. Or, as old Mrs. McGinley used to say, "No one eats a stale crust of bread in a restaurant when there's strawberry shortcake at home."

Assignment 25

By now, your skills in collaboration and in behavioral harmony probably have increased at least a little.

However, you are human, and from time to time you probably have temper explosions. Temper explosions and the resulting fights are not dangerous unless they provoke conflicts—interactions in each spouse fight to defeat the other, to win at the expense of the other. These conflicts, without exception, are deadly to the relationship. There never is one winner and one loser in marriage. There are either two winners or two losers. The mark of the successful relationship is that, regardless of circumstances and differences of opinion, both spouses desire that there be two winners—that the argument ends with both spouses having satisfaction. Successful spouses invest their energies in ending arguments, not winning them.

Conflicts (the destructive interactions in which spouses struggle to defeat each other) are different from problems or arguments. Problems and arguments are behavioral questions set forth by spouses for an orderly solution. All couples set forth and solve scores of problems daily. "Who will meet Julie's bus?" "Who gets to use the bathroom first?"

When orderly problem-solving fails and the subject is of importance to one or both spouses, conflict may result. In this assignment you will learn how to diminish the violence of emotional explosions, or conflict, how to reduce a personal hurricane to a gentle breeze.

THE ASSIGNMENTS

1. Study "Editing Your Negative Thoughts and Feelings before Expressing Them in Observable Behavior."

147

2. Prepare for the "Your Annoying Behavior" exercise.

3. Continue carrying out your Performance Pact.

4. Exchange red and white beans.

5. Continue Intimate Time.

Editing Your Negative Thoughts
and Feelings Before Expressing Them
in Observable Behavior

It is impossible to control the flow of our thoughts and feelings. They drift through our perceptions whether or not we approve. But we can control our *observable* behavior. To control it requires us first to be aware of what we are thinking or feeling.

Once we are conscious of what we are thinking and feeling, we are able to decide whether or not it is appropriate to communicate our thoughts and feelings *in their present form* to other people.

We manage to do this many times a day with strangers and associates in our work. However, with our spouses we too often are inclined to present everything which comes into our heads —especially the negative, destructive thoughts and feelings.

To create and maintain constructive behavioral exchanges with our spouses, we must learn to delay for a moment before projecting our thoughts. This is an old concept. Buddha, 2,500 years ago, advised: "Before speaking or acting toward another person, ask yourself: 'This which I am about to say or do (1) Is it true? (2) Is it kind? (3) Is it constructive to the relationship with the other person?' Unless the answer to all three questions is yes, remain quiet and inactive."

Becoming fully conscious of what we do—*before we do it*— demands considerable self-training and patience. Why? Because being fully conscious of what we think and feel before we transform it into observable behavior is usually contrary to the way we have been conditioned to behave.

In this assignment we will suggest a practical method of delaying the act of automatically projecting negative thoughts and feelings into external actions.

The "conscious delay" will help you take charge of yourself

before sending messages of fear, suspicion, annoyance, dislike. The delay can make it possible, instead, to behave "as if"—to define your spouse positively, to take positive risks, and to exhibit positive expectations.

The delay will give you the chance to behave in *the way you want to behave,* not the way in which your old conditioned habits lead you to behave.

There are those who believe that one always should be "honest and natural," and that behaving differently from the way one thinks and feels is gross hypocrisy. They believe that suppressing negative thoughts and feelings may be psychologically damaging, and that positive procedures such as those we suggest are Pollyannish dreamings which cannot work in everyday married life.

This brings up an often-asked question. Suppose that, in the past, you have frequently *not* gotten what you wanted from your spouse, that he/she has repeatedly ruined certain experiences within the relationship. *Is it still possible to edit negative thoughts and, instead of projecting them, behave in accordance with the law of positive expectations?* (Never mind what has happened in the past. We are not interested in the past.)

The answer is that it *is* possible, and it is imperative if the relationship is to become satisfying.

It is neither hypocritical nor dishonest to edit your thoughts and feelings before you project them willy-nilly into behavior. Only idiots or the irresponsible say everything they think and feel. It is wise indeed to edit negative thoughts and feelings, and to revise them into a constructive "script" before speaking or taking action.

Furthermore, you are not suppressing feelings or energies when you are *aware* of them and *control* them. Quite the contrary! You are converting them—alchemizing them—into the useful information and energies which support a good life as well as a satisfying relationship.

We realize that what we are suggesting requires a dramatic change in your behavior patterns. It requires changing oneself from being a behavioral automaton into a fully aware person— a person who is relatively free from past negative conditioning.

Preparation For The "Your Annoying Behavior" Exercise

Preparation for this exercise begins with each spouse listing those behaviors of the other spouse which provoke anger.

Here are some examples of anger-provoking behavior:

1. S/he is frequently late. When s/he keeps me waiting more than ten minutes, I become angry.

2. S/he embarrasses me before others even though s/he has intended his/her remarks to be good-natured jokes—like describing and imitating my snoring.

3. S/he often overdraws our checking account and doesn't warn me.

4. When I'm busy, s/he sometimes asks me to do measly little chores—things s/he could do as easily as I.

Here are some situations which might provoke anger:

1. Alan asks Barbara to make a business call for him one morning because his schedule is very busy. She refuses because she, too, is rushed.

2. Millie would like to go to the ballet for their anniversary; Sam has already planned to go to the warm-up round of his bowling league that night.

3. John would like to go to his company's Christmas party alone this year. Aileen is immediately suspicious and insists on going along, even though last year she said that she would never attend again.

4. Warren makes dinner Wednesday, Thursday, and Friday nights because Ruth works until seven on those days. He has heated TV dinners for most of the past eleven meals, and Ruth says that she would like to have better food and more variety. Warren tells her to make the meals herself if she doesn't like what he prepares.

5. Every Christmas since their marriage, Dan and Julie have invited Dan's parents and his great-aunt for Christmas Eve and Christmas Day. This year Julie would like to have Christmas alone with Dan and their children. She wants to invite Dan's parents and great-aunt for New Year's Day instead. Dan refuses to ask his parents to make the change, accusing Julie of being selfish.

6. Max and Ellen enter an antique store just to brouse. Max admires an antique table and starts a conversation with the shop-owner about it. Anxious to go on to the next shop, Ellen interrupts by commenting that Max shouldn't take up the shop-owner's time because he really doesn't understand antiques.

Now that you have read the examples, each of you write a description of one of your spouse's anger-provoking behaviors. Choose one of the most dramatic situations/behaviors possible. By "dramatic" we mean situations and behaviors which realistically can be acted out by you and your spouse. In tomorrow's assignment that is what you will do—perform a mini-drama.

To assist you in choosing the most dramatic situation, you may read the next assignment, which describes how the mini-dramas are acted out.

Assignment 26

In today's assignment both spouses will star in two, one-act plays—the re-enactment of the anger-provoking situation/behaviors which each has chosen.

It may appear strange to suggest that you act out a negative behavior. There are two reasons: One, you will sense the absurdity of such explosive behavior. Second, later you will practice a method of avoiding such behavior in the future.

THE ASSIGNMENTS

1. Do the "Your Annoying Behavior" exercise.

2. Continue carrying out your Performance Pact.

3. Exchange red and white beans.

4. Continue Intimate Time.

The "Your Annoying Behavior" Exercise

In today's exercise, both spouses will star in two dramatic productions entitled "Your Damned Annoying Behavior". Besides starring in both productions, each spouse, in turn, also will be the stage director.

The director is in charge of the drama which illustrates the situation/behavior which annoys him/her. S/he sets the scene for it, describing where the scene takes place and under what circumstances. Then the two stars—wife and husband—act out the situation, the resulting temper-provoking behaviors, and the almost inevitable conflict which follows.

Act it up! Get as much anger, emotion, conflict, as possible from the scene. Make it as real as you can.

Flip a coin to find out who is the first stage director. In our example, John wins the toss. He has chosen the following situation/behavior: *Mary is frequently late. When she keeps me waiting more than ten minutes, I lose my temper.*

Since John has won the toss to be the first director, here are John's stage directions to Mary.

SETTING THE SCENE
"Mary, the living room represents the corner of Fifteenth Street and Park Avenue, where you are supposed to pick me up at exactly ten o'clock at night. I got there five minutes early. It is January. The temperature is below freezing, and the wind is blowing. I'm cold and hungry. At ten o'clock you're not there. You finally show up at twenty-five minutes after ten. I'm freezing and angry.

"Okay, Mary, that's the scene."

DESCRIBING THE ACTION
"Mary, you go outside the room and then come back in, your hands on an imaginary steering wheel, driving an imaginary car. You pull up to me—me, who's freezing on the streetcorner.

You open the car door, and before I can say anything, you start talking fast and loud—knowing I'm in a rage. Ad lib everything you say. I'll do likewise, and we'll keep at it, getting as nasty and angry as possible."

ACTING OUT THE SITUATION

Mary leaves the room. She re-enters, "driving a car." John is flailing his arms to keep warm, puffing, stamping his feet. Mary drives up to John, opens the imaginary car door, and says, "Darling, I'm sorry I'm late."

John gets into the imaginary car (Mary and John sit on chairs next to each other).

John shouts: "Damn it, you're late, as usual. I'm here freezing my butt off—and where were you? I suppose belting in the booze with some of your kooky lib friends. Why the hell can't you be on time once in a while? Especially on a night when it's about twenty below!"

"Look who's talking! I do you a favor to pick you up—which screwed up my plans for the evening. Then I had problems you don't even know about. The car. . . ."

"I told you to get the damn car fixed, but you knew better."

"Don't ever ask me to pick you up again. Next time take a taxi!"

By now John is pretending (or is he?) extreme anger: "If you weren't so careless with money, I'd be able to afford a taxi and wouldn't have to depend on a selfish, irresponsible person like you."

"John, you're sick. You'd get a promotion and make more money if you weren't so paranoid and if you had the guts to be a man."

"Damn you! I'll take a taxi home rather than ride with a foul-mouth like you. And I'll use the house money to pay for the taxi."

John opens the door and jumps out.

That's the way to do it. Let it all out. Ad lib your way through the scene as if it really were happening and developing into a runaway conflict.

There will be two one-act plays. In the second one, the other spouse (in this case, Mary) is the one who directs the play—because it involves *her* anger. When both of these prize-winning performances are over, shake hands and congratulate each other on the marvelous directing and acting.

Assignment 27

*Yesterday you acted out two scenes concerning anger-
provoking behavior which, if unchecked, could lead to conflict. In today's assignment you will perform the same two
one-act plays which you did yesterday. However, there will
be a radical change in stage direction: the two plays will
be acted in a way which will illustrate how to delay and
dilute a conflict. This does not mean that you contain the
anger within you; rather, you transform it into a more
constructive, a more positive, energy. When this is accomplished, it is likely that the potential for conflict will be
diminished; and that the compassionate intimacy between
the spouses will be increased.*

THE ASSIGNMENT

Act out the two plays, first the negative version, then the
positive one, which shows how to delay the behavioral explosion.

Diminishing the Behavioral Explosion

Today you will start with the same scene as yesterday. However, today, before your acted-out anger approaches the point of conflict, stop the action. In this interval, which delays the conflict eruption, ask yourself four peace-bringing questions:

1. What are some of the possible explanations for my spouse's anger-provoking behavior?

2. Have I the right to attack my spouse on this issue? Am I blameless? Have I been constructive and positive in recent interactions?

3. If I assume even the worst possible explanation for my spouse's behavior, will it help me if I cause conflict over it? Will a bitter fight help our marriage? Will it reduce the chances of my spouse repeating the same behavior?

4. Having answered questions 1, 2, and 3, do I want to initiate a conflict at all?

We will start with the same scene as in the last example. Mary was supposed to pick up John at the corner of 15th and Park at ten o'clock. It is a cold, blustery night. Mary is twenty-five minutes late. John, in a temper, is on the corner, flailing his arms, shivering, stamping his feet, and cursing.

As she did yesterday, Mary goes out of the room and then returns, "driving the car," and pulls up to the angry, freezing John. She opens the door and says, "Darling, I'm sorry I'm late."

Now the new stage directions:

Before John explodes emotionally and viciously attacks Mary, he asks himself—out loud, so that Mary can hear him—peace-bringing question number one: "What are some possible explanations for Mary's lateness?"

The dramatic action is temporarily suspended. The action in the scene "freezes" while John is thinking out loud.

Having asked himself the question, "What are some possible explanations for Mary's lateness?" he answers his own questions —aloud.

"Perhaps the car broke down or she had a flat tire."

"Perhaps I gave her wrong directions. I usually ask her to meet me at the office—perhaps I didn't make it clear that it was to be at Fifteenth and Park."

"Perhaps something important came up and put her behind schedule."

"Perhaps she was trying to get even with me because at dinner last night I criticized her publicly."

The action is still frozen. John says to himself (still out loud),

"Certainly there is no sense in my getting angry if the car had a flat or if I gave Mary wrong directions."

John then asks (again, out loud) question number two (which really is a series of questions): "Have I the right to scold Mary now, to go into battle to put her in her place? Am I blameless on this particular issue? Have I, in recent interactions with Mary, generally been constructive and positive?"

He answers the multiple question out loud: "I'm not sure I have the prerogative of attacking Mary. Last week I was late for dinner twice, and I must admit that I've been doing some things I know she doesn't like."

Next he asks himself question number three: "If I assume the worst about Mary and her motivation, what will I gain if I battle her over it? Will it help me? Will it help our marriage? Will it prevent Mary from being late in the future?"

John answers his own question aloud.

"Well, I can take my rage out on her. However, if Mary is innocent of what I blame her for, then she, in turn, will be angry and may strike back. Then we'll have another two- or three-day battle."

John continues answering his own questions: "If Mary's intentions were mean and bad, then my anger would give her a perverse satisfaction. She would know that she can make me lose my temper whenever she wants. She would know that she can manipulate me."

He continues out loud: "If I threaten her or humiliate her, she may retaliate either now or later."

He thinks a bit more: "No, I'd better withhold my explosion of anger."

Having considered the first three peace-bringing questions, John now asks question number four: "Do I want to initiate a conflict with Mary at all?"

John answers himself out loud: "If I delay, perhaps once I cool off we can discuss this constructively."

Having gone through the four peace-bringing questions and the delay, John unfreezes the scene and directs that the play continue.

Mary opens the door of the imaginary car again and says, "Darling, I'm sorry I'm late."

John, getting into the car (and trying to transform the possible conflict into a positive interaction), responds:

"Whew! Mary, it's cold out there. I'm frozen, and let me tell you, I'm sure glad to see you."

"I'm sorry about it, John. I really am. The car wouldn't start. I got Mrs. Smith to jump-start the car for me, and then she started telling me her problems. I guess I should have told her you were waiting in the cold and offered to see her later."

It is easy to see that the possible conflict can be transformed into a positive interaction, and the play probably will have a reasonably happy and positive ending. In the majority of cases, people who go through this process of the delay and the peace-bringing questions are able to avoid conflict. Furthermore the process eliminates the I-am-right-and-you-are-wrong attitude and promotes real intimacy, trust, and equality between the spouses.

Now the other spouse (in this case, Mary) who directs the play involving her anger and goes through the same procedure of asking *aloud* the four peace-bringing questions.

P.S. In our clinical experience, most spouses find this exercise so satisfying that they voluntarily continue acting out several more annoying behavior scenes.

In follow-ups on our clients, we have witnessed spouses approaching an explosion. Frequently the situation softens as one spouse smiles, touches the other, and says, "Let's ask the questions and act it out."

Assignment 28

This assignment and the two which follow are probably the most difficult in the book. However, for couples whose marriage is troubled by quarreling, they are probably the most important. To get through these three assignments will require all the skills you have learned so far. More than that, considerable patience and stamina are needed in studying the unavoidably complex instructions.

These assignments concern the equitable allocation of power and authority within the marriage.

One definition of power is "the possession of control, authority, or influence over others."

The Oxford English Dictionary *defines* authority *as "power or right to enforce obedience . . . the right to command or give an ultimate decision."*

In a marriage, when one spouse has the preponderance of authority or power, then s/he has the preponderance of control (over the other spouse) and makes the preponderance of decisions on the basic activities within the relationship. When this occurs, the other spouse feels inferior, dependent, downtrodden, abused, and will do almost anything, *including divorce, desertion, or even physical violence, to escape the situation.*

Therefore, when spouses are unable or unwilling to agree on an equitable allocation of authority and power, a power struggle is inevitable. If this conflict is not resolved, it is impossible to enjoy a state of satisfaction.

This assignment—along with the next two—suggests how to negotiate an equitable allocation of power and authority between the two spouses.

THE ASSIGNMENTS

1. Study "The Anatomy of Authority and Power Allocation."

2. Continue carrying out your Performance Pact.

3. Continue Intimate Time.

The Anatomy of Authority and Power Allocation (Who Is in Charge of What)

Within the marriage there are approximately twenty areas of activity which comprise the core of the relationship. The way these activities are carried out defines the couple's social structure. The way these activities are carried out requires decision-making, and the decision-making is accomplished via the *authority and power behaviors.*

The areas of authority and power are as follows (the designating letters in the left column will be explained later. Disregard them at this time):

The control of:

A   What work, if any, the wife does outside of the home (business)

B   What work, if any, the husband does outside of the home (business)

C   The budgeting of the total income

D   Who spends the budgeted money (makes purchases, pays bills, keeps accounts)

E   Where the family lives (what geographical area as well as what neighborhood and type of house)

F   How to arrange and furnish the house (allocating space, decorating, furniture, appliances)

G   What work the husband does at home (cooking, household chores, child-rearing, home repairs, etc.)

H   What work the wife does at home (cooking, household chores, child-rearing, home repairs, etc.)

I   When and how to have sex

J   Who uses the birth control, and which method is used

K How many children to have

L When and how to praise or punish the children

M The time and activities of the family as a group

N The time and activities of the spouses together, exclusive of children

O The time and activities of the spouses apart from each other (clubs, social events, hobbies, sports, etc.)

P How and when to entertain friends

Q How and when to entertain members of the wife's family

R How and when to entertain members of the husband's family

S Whether to attend church, which church, when

T When and where to spend vacations

U, V, W Others which are unique to your relationship

The twenty or so activity areas listed above are realities in almost all families. They are activities on which most spouses have strong feelings; and if the spouses have stubbornly-held differences of opinion on how power in these activities should be manifested, conflict is inevitable.

Unresolved disputes over "who is in charge of what" is one of the most powerful and prevalent causes of conflicts and divorces.

To avoid this kind of conflict over "who has the authority and power in each of the twenty or so core activities," the spouses must objectively (not emotionally) become aware precisely of "who wants what." When this identification has been completed for the entire spectrum of marital activities, then the spouses must negotiate and reach agreement on the following questions:

1. *Who will decide* what action is to be taken within each of the authority and power areas?

2. *When* and *how* will the action be taken?

3. *Who* will take the action?

4. *Who* ordains the rules which govern decision-making for the basic authority—and power—behavior.

The process of accomplishing the above goals is the Authority and Power Allocation exercise which you will now are begin. The exercise is accomplished in three steps:

STEP ONE (TO BE COMPLETED IN THIS ASSIGNMENT)

Each spouse defines and describes, how s/he currently perceives the authority and power structure of the relationship, or, rephrasing it, "who currently is in charge of what, and to what degree?" This assists spouses to become objectively aware of how each feels the marriage is being run now. *In the majority of unhappy marriages, the spouses are unaware of how much power each has; and emotionally believes the other spouse has far more power than s/he deserves.*

STEP TWO (TO BE DONE IN ASSIGNMENT 30)

Each spouse defines and describes, the way s/he would like the authority and power to be allocated in each of the twenty or so core activities of the relationship. This objectively informs each spouse precisely what power and authority the other spouse wants. *In the majority of unhappy marriages, the spouses are unaware of what either of them wants—even though there are quarrels over frustrated wants.*

STEP THREE (TO BE DONE IN ASSIGNMENT 31)

The spouses negotiate about "who is in charge of what" within the activities over which there is disagreement. The negotiating continues until both spouses feel comfortable and satisfied with the compromises which have been worked out.

We now will start off with Step One: Each spouse defines and describes how s/he currently perceives the authority and power structure of the relationship.

STEP ONE

Because of the complexity of this exercise, we will start with an example. The first of the following two charts (one for the wife and one for the husband) shows how two spouses each perceived the allocation of authority and power within their troubled marriage.

We will follow this couple through the entire three steps of the authority-and-power re-allocation process. The example is from a case history of a couple who had deep problems but wanted to stay married. As in many troubled marriages, there is much sexism in the case history. The battle of the sexes and the resulting difficulties are among the most prevalent and most complex to work out. That is why this particular case history has been chosen.

The Case History

The wife and husband began Step One by looking over the twenty or so areas of authority and power listed at the beginning of this assignment. (For convenience, the list is repeated on each of the charts below.)

The wife and husband indicated on Charts 1 and 2—by placing the designating letters within the appropriate boxes—how each believed the authority and power was currently distributed within the marriage; who was in charge of what; and how much each is in charge.

EXPLANATION OF THE CHARTS

In our example, the wife believed that Item A (what work the wife does outside the home)—was mostly controlled by her husband; that is, she believed that he made the decision after discussing the situation with her, and that she had little influence in this area. Therefore, she put the letter *A* in Box 2 of Chart 1 (the how-the-wife-perceives-it chart).

The husband, however, perceived things differently. He believed that the wife made the decisions concerning her work outside the home, after consulting him—in other words, that she had more power in this area than he did. So he put the letter *A* in Box 4 of Chart 2 (the how-the-husband-perceived-it chart).

CHART 1: HOW THE WIFE PERCEIVES THE PRESENT ALLOCATION
OF AUTHORITY

husband has absolute authority and makes decisions without wife's approval	husband makes decisions with wife's consultation and approval	wife and husband make decisions together	wife makes decisions with husband's consultation and approval	wife has absolute authority and makes decisions without husband's approval
B C G	A D T	I N R K O S E P H M Q		J L F
Box 1	Box 2	Box 3	Box 4	Box 5

A   Wife's work outside home
B   Husband's outside work
C   Budgeting total income
D   Spending budgeted money
E   Where does family live
F   How to arrange and furnish house
G   Work husband does at home
H   Work wife does at home
I   When and how have sex
J   Birth control (who and how)
K   How many children to have
L   Praising or punishing children
M   Activities of family as group

N   Activities of spouses together, exclusive of children
O   Activities of spouses apart from each other
P   How and when to entertain friends
Q   How and when to entertain members of wife's family
R   How and when to entertain members of husband's family
S   When to attend church, which church, and when
T   When and where to take vacations
U, V   Others which are unique to your relationship

CHART 2: HOW THE HUSBAND PERCEIVES THE PRESENT
ALLOCATION OF POWER AND AUTHORITY

Husband has absolute authority and makes decisions without wife's approval	Husband makes decisions with wife's consultation and approval	Wife and husband make decisions together	Wife makes decisions with husband's consultation and approval	Wife has absolute authority and makes decisions without husband's approval
	B D T	K N C P E R M Q	A O	I L J F
Box 1	Box 2	Box 3	Box 4	Box 5

A Wife's work outside home
B Husband's outside work
C Budgeting total income
D Spending budgeted money
E Where does family live
F How to arrange and furnish house
G Work husband does at home
H Work wife does at home
I When and how have sex
J Birth control (who and how)
K How many children to have
L Praising or punishing children
M Activities of family as group

N Activities of spouses together, exclusive of children
O Activities of spouses apart from each other
P How and when to entertain friends
Q How and when to entertain members of wife's family
R How and when to entertain members of husband's family
S When attend church, which church, how, and when
T When and where to take vacations
U, V Others which are unique to your relationship

For Item B (what work, if any, the husband does outside of home—his business or profession), the wife believed that the husband had complete authority and power and that he made decisions concerning his job without consulting her. Therefore, she put a *B* in Box 1 of Chart 1 (the how-the-wife-perceives-it chart).

The husband, however, believed that he made decisions about his business only after consulting his wife, so he put a B in Box 2 of Chart 2.

For Item C (budgeting of total income), the wife believed that her husband had absolute authority in this area and that he made decisions without her approval. She put a *C* in Box 1 of Chart 1.

The husband perceived things differently; he believed that the budget was decided on by his wife and him jointly, as equals. He put a *C* in Box 3 of Chart 2.

Item I concerns when and how to have sex. The wife believed that she and her husband jointly made these decisions, so she put an *I* in Box 3 of Chart 1. The husband perceived his wife as having complete decision-making authority when and how to have sex. He believed it was her changing moods and attitudes which controlled "when and how." Therefore, he put a *I* in Box 5 of Chart 2.

In this manner the wife and husband went through the entire list of activities. Charts 1 and 2 show the completed process.

Now we want the two of you to make out your first charts.

There is no need to duplicate the entire chart. It is only necessary to draw two sets of five boxes:

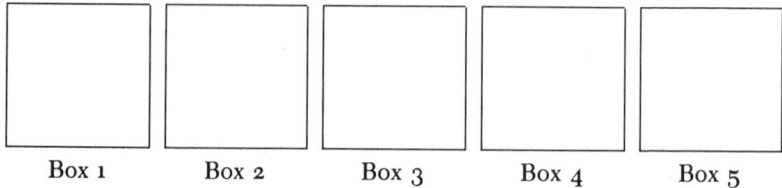

Box 1   Box 2   Box 3   Box 4   Box 5

One set of boxes is for the wife and one for the husband.

Each indicate on her/his chart how s/he believes the powers and authority in their marriage currently are being controlled. When the boxes have been filled in as explained above, look at each other's charts. It is probable that, in several areas of activity, there will be differences of opinion about who has the authority and power. This surprises some couples because each is certain that it is obvious who is in charge.

We urge you not to discuss who is right and who is wrong. Your goal is to reach a solution in which both of you are "right" and satisfied. In tomorrow's assignment you will let each other know how you would like to see things changed.

Assignment 29

In this assignment each of you will indicate whom you want to have in charge of each area of activity—and how much in charge.

THE ASSIGNMENTS:

1. Fill in Charts 3 and 4 showing how each spouse would like to see the authority and power re-allocated.

2. Continue carrying out your Performance Pacts.

3. Continue Intimate Time.

The Way You Would Like to Have Authority and Power Allocated

Each spouse now draws five more boxes. In our example, for your convenience, we will use the complete charts (3 and 4) with explanations as to what the boxes and the letters mean.

Go down the list of authority and power areas (A, B, C, etc.). Put the letters in the boxes *according to the way you would like* to have the decision-making in the various areas carried out.

For example, in the last assignment (how things are now) the husband had *A* in Box 4 of Chart 2 (meaning that his wife, after consulting him, made decisions concerning her job outside the home). However, he would prefer that this be a jointly made decision in which she and he have equal authority and power. Therefore, in the present assignment he moved *A* into Box 3 of Chart 4.

In the last assignment, the wife believed that her husband made decisions concerning her outside job, after consulting her, so she placed *A* in Box 2 on Chart 1. However, *she* wanted to make these decisions after consulting her husband. Therefore, in this assignment, she placed *A* in Box 4 of Chart 4, indicating that she wanted more authority and power in that area.

Charts 3 and 4 show how the wife and husband in our case history *wanted* the authority and power allocated.

Listing Agreements and Disagreements

By examining the Charts 3 and 4 (the how-I-want-it charts), we see that the husband and wife were in agreement on Items B, K, E, M, N, O, P, R, Q, H, and G. These are the items which both spouses placed in the same numbered boxes.

However, they disagreed on Items A, I, J, C, D, F, L and T.

CHART 3: HOW THE WIFE WANTS AUTHORITY AND POWER
ALLOCATED

Husband has absolute authority and makes decisions without wife's approval	Husband makes decisions with wife's consultation and approval	Wife and husband make decisions together	Wife makes decisions with husband's consultation and approval	Wife has absolute authority and makes decisions without husband's approval
	B	K M Q C N T D O S E P H L R G	A I	J F
Box 1	Box 2	Box 3	Box 4	Box 5

A   Wife's work outside home
B   Husband's outside work
C   Budgeting of total income
D   Spending budgeted money
E   Where does family live
F   How to arrange and furnish house
G   Work husband does at home
H   Work wife does at home
I   When and how have sex
J   Birth control (who and how)
K   How many children to have
L   Praising or punishing children
M   Activities of family as group

N   Activities of spouses together, exclusive of children
O   Activities of spouses apart from each other
P   How and when to entertain friends
Q   How and when to entertain members of wife's family
R   How and when entertain to members of husband's family
S   When attend church, which church, and when
T   When and where to take vacations
U, V   Others which are unique to your relationship

CHART 4: HOW THE HUSBAND WANTS AUTHORITY AND POWER
ALLOCATED

Husband has absolute authority and makes decisions without wife's approval	Husband makes decisions with wife's consultation and approval	Wife and husband make decisions together	Wife makes decisions with husband's consultation and approval	Wife has absolute authority and makes decisions without husband's approval
	B D T C	A E O S I F P H J M R G K N Q		L
Box 1	Box 2	Box 3	Box 4	Box 5

A Wife's work outside home
B Husband's outside work
C Budgeting total income
D Spending budgeted money
E Where does family live
F How to arrange and furnish house
G Work husband does at home
H Work wife does at home
I When and how have sex
J Birth control (who and how)
K How many children to have
L Praising or punishing children
M Activities of family as group

N Activities of spouses together exclusive of children
O Activities of spouses apart from each other
P How and when to entertain friends
Q How and when to entertain members of wife's family
R How and when to entertain members of husband's family
S When attend church, which church, and when
T When and where to take vacations
U, V Others which are unique to your relationship

Your Lists of Disagreements

Working together, make your own list of the items on which you disagree. It is these items which you will negotiate.

List the disagreements in order of ease of negotiating. That is, at the top put the subject which you both believe can be agreed upon most easily. If possible, this should also be the subject on which you feel most generous, on which you would willingly give the advantage to your spouse.

You may not agree on the order of priority. In that case, flip a coin to see who makes the choice of the top subject, and from then on alternate in choosing the order. With Mary and John, the priorities were:

L   Praising and punishing children
F   Arranging and furnishing home
T   When and how to take vacations
C   Budgeting total income
D   Spending budgeted money
A   What work, if any, wife does outside of home
I   How and when to have sex
J   Who uses birth control, and which method is used

When you have made out your list, then you are *almost* ready to negotiate.

Assignment 30

Getting an Overall View Before the Power Negotiations Begin

Some spouses become so obsessed with one or two issues in the marriage that they are unable to consider others. In this assignment we suggest a method for acquiring a balanced, overall view of everything to be negotiated.

THE ASSIGNMENTS

1. Read "The Overall Approach."

2. Study an example of how Mary and John worked out an over-view of their situation.

3. Make a relationship summary of your situation, patterned on the example of John and Mary.

4. Continue carrying out your Performance Pact.

5. Continue Intimate Time.

The Overall Approach

Disagreements concerning power usually involve several areas of marital activity. In the example of Mary and John, there was disagreement in the areas of child rearing, vacations, home management, finances, outside jobs, and sex.

The inclination of negotiating spouses is to work on one disagreement at a time, while keeping all others on the shelf. This has the advantage of making it easy to keep one issue in focus. But, it has the serious disadvantage of forcing decisions on single issues which are out of context with other pressing concerns. The relationship is a system of many issues, and each subject has a bearing and influence on all others.

The "total approach" requires that spouses keep many different subjects simultaneously in focus. This enables them to see how the condition of one marital activity influences all others. This also makes it possible for spouses to be conscious of the total inventory of good things which they can offer to their spouses as compromises or concessions.

Therefore, before negotiating, it is necessary to grasp the overall picture of the relationship. This can be accomplished by discussing the family's total situation, *focusing on the areas of disagreement.* The example which follows will make this clear.

Some couples find it easy to write about their family situation. Others prefer to talk about it. We suggest writing it *after* talking. The résumé of the situation should be done objectively, *as if it were a description of two other people.*

This what you do in today's assignment.

Here is the résumé of Mary and John. (They discussed it and then wrote it together, using the third person form to help in retaining objectivity.)

They have three small children, ages 2, 4, and 5. John's income is barely enough for their family to live on, with little left over for luxuries. Mary looks after the cooking and runs the home. She also operates a modest home industry, designing and making children's clothes which neighborhood stores sell. She nets about $150 monthly. She spends most of this money on her personal needs—beauty parlor, cosmetics, clothes, and so forth. She has a great fear of poverty; before her marriage she worked hard and made enough money to live very well. Mary is a capable clothing designer, a superb seamstress, and an excellent business manager. She is frustrated because the requirements of three small children rule out a full-time, money-making job. She sews mostly at night and on Sundays, when John takes complete charge of the children. She makes the rounds of the stores on a catch-as-catch-can basis on weekday mornings. Her dream is to have a full-time business of her own, but together John and Mary do not have the capital to start it.

Mary believes that the only way she can raise the capital and also help John out so that they can live better is by taking a night job outside the home. She has been offered a job as a cocktail waitress in a restaurant-nightclub from 6:00 P.M. to 1:00 A.M. She could make about $300 a week in salary and considerably more in tips, if she is a good waitress. Mary believes that this job would solve most of their present economic problems and that in several years she could accumulate enough capital to start her business. She is determined to take the job.

If she takes the job, John will have to prepare breakfast and dinner for himself and the children, and look after the children six nights of the week.

John does not want Mary to take the job—not only because he does not want to be tied down and lose overtime of his own, but also for another reason which is important to him: He knows that Mary is attractive to other men, and he does not like the idea of her wearing a scanty outfit in what is primarily a men's bar, staying up late in the company of men who are drinking, and coming home late at night alone.

The issues of the job and money concern areas of behavior over which Mary and John disagree violently.

Also, there is disagreement over the issues of sex and children. Mary wants no more children. However, she does not want to take the pill because she is afraid it may cause cancer. The diaphragm and the IUD irritate her; and she is not confident of their reliability. The rhythm method is not sure enough for her. She wants John to have a vasectomy. However, John is afraid of this. He does not know what it will do to him psychologically and physiologically. He also feels that at some future date Mary and he might want another child.

Mary will have sex with John only when it is "safe" according to the rhythm method, and only if, at the same time, John wears a condom. But even then she is nervous and inhibited.

Before they had children (and Mary was working full time at a high-paying job) Mary and John had a highly successful sex life.

John and Mary also disagree on who should make the family budget and who should spend the budgeted moneys. John believes in paying cash for everything; if they do not have the cash, he thinks they should do without. Mary disagrees. She believes they should buy such items as a new car, a color TV, a vacation and so forth on credit and pay the bills over a long period of time.

John wants to control of the spending of the family's money. He is annoyed because Mary spends her home-industry earnings on herself and thereby she has luxuries which the family budget does not allow him to have. He says that if she put her earnings into the family pot, he would be glad to have Mary share in decision-making about the budget.

John argues that if they can endure things the way they are now, without Mary taking the cocktail waitress job, then, within three years, all the children will be in school, he will be making considerably more money, and Mary can then work full time during the day and accumulate money to start her own clothing business.

Mary says that she does not want to live for another three years under the present conditions.

Mary is dissatisfied with their annual vacation. Every summer the family goes to a hunting lodge owned by John's brother. It is free, but Mary does not like it because it is primitive. Mary feels that going there may be a vacation for John, but it is not for her and the children. She would prefer to go to a resort where babysitters are available and there are luxuries which they do not have at home. She is willing to go into debt for a vacation.

Now It's Your Turn

After having studied the relationship summary of Mary and John, make up your own. Discuss it. Consult the power charts you have made. Then write out your own summary. If there are older children, have them participate. They are as influential in determining the family system as you, the parents, are. You cannot keep secrets of the relationship from them, anyway. *The children know what's going on.* If you leave them out of the process of changing the system, the job becomes far more difficult.

Assignment 31

In this assignment you will read about how a couple with fairly deep and complex disagreements concerning authority and power went about negotiating their disagreements. Also in this assignment—because you will begin your own negotiating tomorrow—are a few reminders of how to behave while negotiating.

THE ASSIGNMENTS

1. Study "The Negotiations of John and Mary."

2. Study "Some Things to Remember Before Your Actual Negotiations Start."

3. Continue carrying out your Performance Pact.

4. Continue Intimate Time.

Preparing for the Negotiations
(How John and Mary Did It)

Draw the Final Authority and Power Chart. Make it now even though you won't need it until tomorrow. It is on this chart that wife and husband will indicate the way they *both* agree the power should be allocated. Therefore there is only one chart.

1	2	3	4	5
Husband makes decision without consulting wife	Husband makes decision after consulting wife	Both husband and wife make decision together	Wife makes decision after consulting husband	Wife makes decision without consulting husband

The discussion should begin with the issues which appear easiest to solve, and on which spouses can be most generous.

The spouse who begins the negotiations is the one who says s/he has a compromise, a generosity, an idea which will start things off in a way which will benefit the relationship, an offer which will promote a "two winner" model.

If both spouses say they have a gift or a generosity with which to start they flip a coin to see who starts.

In our example, John won the toss. However *he did not* start with the first items on their disagreements list. Their list was:

L Praising and punishing children
F Arranging and furnishing home

T   When and how to spend vacations
C   Budgeting total income
D   Who spends the budgeted monies
A   What work, if any, wife does outside of home
I   When and how to have sex
J   Who does birth control and how

John, instead of starting with Item L, began with Items C and D. He did this because, after thinking things over, he decided that Mary had deep feelings on the subject of budgeting and spending, and that these subjects influenced Mary's thinking on other activities in the marriage. John felt that Mary's concerns on Items C and D were justified and that the difference was very small. He concluded that giving in to Mary on these items was something on which he should be generous.

Therefore he told Mary he thought it would benefit the marriage if he and Mary made out the budget together every month (Item C) and that at same time they should jointly decide who would spend the budgeted monies (Item D), for example, Mary would take care of the food budget, John would take care of the utilities, etc. (Item D).

Mary liked his suggestion and agreed. She responded with a generosity of her own. She said she would put half of any money she made into the family general fund—to be included in figuring out the budget. They decided to try this for two months. At the end of two months they would negotiate this item again. If the system worked they would leave it as is. If there were problems, they would attempt to straighten them out at the renegotiations in two months.

Mary and John put the letters for their agreed-upon items C and D into Box 3. Next to each letter they put the notation "2 months."

Mary took the next subject. She said she thought it would be good for the family if John participated more in praising and/or punishing the children. She volunteered to have dinner earlier so that she and John could be with the children a half hour every evening before the children went to bed. During this period of family gathering, she could review what the children had been doing during the day; and then John could join her in praise or discussion, as appropriate. Mary acknowledged that this would be difficult if and when her cocktail waitress job went into effect, which could not possibly be for a month. Therefore she suggested a one-month trial before renegotiation.

They placed the letter L into Box 4. This meant the decisions about praising and punishing the children will be largely Mary's, but this will be done after John and she have consulted and John has approved. After the letter L they added the notation "1 month."

John said he understood how Mary felt about vacationing at his brother's hunting camp in the mountains. He said that Mary sometimes had mentioned the possibility of their staying at her mother's house which is on the ocean about a thousand miles away. Mary's mother would be glad to see them and act as babysitter part of the time. John's objection had been that driving that distance in their small auto with three infants would be a trying experience; and flying down would require more money than was available. He suggested that his brother might lend them his van, and they could drive down in that. Their vacation started in four months; could they try that for this year? Mary agreed. They put Item T into Box 3 with the notation "5 months."

Mary said that she has been stubborn about decorating the house and allocating space not only because she felt this was traditionally the wife's role, but also because John had been reluctant to help her with room painting and wallpapering— usually saying he thought the house looked okay or that he had other things to do. Mary said that if she could get a *little* help from John, then they could make the decorating decisions together—which would be good for their relationship. John responded that he would help her with the heavy work over weekends; however, he said that because she is at home more than he is, the way the house looked and was arranged was probably more significant to her and that she was more knowledgeable than he on the family's needs. Therefore he would like to consult with her and then, after deciding together, have her be in charge of this activity. Therefore they agreed to put item F in Box 4, and have the agreement last for 2 months before renegotiations.

Mary now brought up the matter of sex, item I and J. She explained that she very much enjoys sex with John; but that she is frightened of having another child; and that because of the trouble she's had with the diaphragm and the IUD, she will only be satisfied and feel secure if John has a vasectomy. She said she understands his apprehension regarding the operation.

Mary said she saw no solution, but that between the two of them *they had to find one.*

John said there is a sex clinic at the university. They see patients and have instruction at night. The course meets once a week for three months. Perhaps if they attended the course together they might find a solution to their mutual problem on sex.

Mary agreed that perhaps the solution could be found that way and agreed they go to the sex clinic for the next three months, providing the clinic does not interfere with her nighttime job if she took it.

They placed the letter I and J in Box 3 tentatively and noted that the duration of agreement is three months. They also agreed that I and J might have to be moved to other boxes after the negotiation on Mary's waitress job, if she took it.

The last point of issue was item A, which concerned Mary's possible job.

Mary and John discussed this and made no progress. They were beginning to get discouraged and angry.

Mary said, "What we're trying to do is find a way for me to have a business of my own, an activity of my own, plus having enough money as a family to live the way we want, plus bringing up the children properly, plus enjoying each other. Perhaps we must seek a solution other than the ones we've been working on.

"If we had the capital, I could start my own business with proper equipment, proper space, and a bit of help. I believe that within a year—with your help, John—we could have a family business which would make us prosperous. But the fact is, we do not have capital; and we don't even have anything on which we can borrow, to use as collateral for a loan. Our house already has maximum mortgage on it. The car is too old. We have no stocks or property."

They discussed this subject, trying to decide how much money was needed. Mary said they could convert the attic into a workshop for about $2000. Her extra equipment would, at the start, come to about $2000. Then about $3000 for immediate operating expenses.

John said he saw no way to get that much money, but let him think about it.

The next day John said he had spoken to the bank which said that on the basis of Mary's "part-time" business over the last year, the bank would lend them $10,000 if they could get a cosigner. John said his brother had agreed to cosign.

Mary and John put Item A into Box 3, with a one-year notation after it. Mary agreed that they could concentrate on expanding her "home industry" in the attic; and after things began prospering she might move to another location, and the new space in the attic could be used either as a den for John or as added bedrooms for the children, this to be decided later.

That ended their negotiations. Mary began building her business, using the $10,000 loan. She did not need her waitress job. At the conclusion of the sex clinic, John felt he would feel secure about having a vasectomy, and he had the operation performed. He has had no negative effects, and once again he and Mary enjoy a good sex life. Mary's business has prospered.

The relationship has become richer and richer in individual and mutual satisfactions.

Every six months, Mary and John still review their Cherishing Days, their Performance Pacts, and their Authority Charts.

You are now almost ready to negotiate your own Authority Behaviors, in the same way Mary and John did. We suggest you take time to discuss Mary and John's case. Then study the short text which follows: *some things to remember before the actual negotiations start.*

Some Things to Remember Before the Actual Negotiations Start

You are now ready to negotiate the authority and power allocation. To do this successfully, you must be aware of several things:

1. The aim of the negotiations is to have a "two-winner" model, in which both spouses feel they will enjoy the maximum gains from the new arrangements.

To accomplish this "two-winner" model, both spouses must avoid coercion or manipulation in what might be a selfish effort to obtain personal gain (instead of mutual gain). If, as a result of coercion or manipulation, one spouse *temporarily* should "win" and the other *temporarily* "lose," the loser ultimately will find a way of getting compensating satisfactions in some other manner. Thus the negotiations will be aborted.

2. During negotiations spouses must eliminate behaviors which distract from the positive qualities of the negotiations. For examples:

Avoid bringing up any unpleasantries about the past.
Avoid bringing up any behaviors or qualities of your spouse which you consider negative or inferior.

You already have learned how to do this. You knew that even though it is impossible to control one's thoughts, still, one can control what s/he says and does. You also know which of your behaviors provoke your spouse. Avoid them at all times, but especially during these negotiations.

3. The spouses constantly must remember that agreements reached by negotiation are not binding for all time. Every

agreement reached is a *tentative solution* which will be re-evaluated and renegotiated within several months. Because the negotiated agreements are binding for only short periods of time, it is easy to take positive risks.

4. Make certain that something is offered by you in exchange for everything you want changed. This will strengthen reciprocity in the relationship as well as speeding up the negotiations.

5. Make only those commitments you are prepared to honor. If you are not willing to abide by agreements, do not make commitments just because you want to gain concessions in other areas. This damages credibility in future negotiations.

6. There will be times when both spouses take firm and incompatible stands, with neither being willing to make concessions.

These are the times to stop and remind each other that there probably is a different path which will bring the desired compromise, and with it a new, successful solution.

Mary and John had two such occasions, and by having a "think tank" session they found mutually agreeable solutions.

7. The effectiveness of your negotiations can be influenced by time, place, and circumstances. There are many conditions which can diminish your clarity of mind, your good will, your desire to negotiate a "two-winner" model. For examples, fatigue can cause irritability; a high noise level forces you to shout at each other, which increases tension; unexpected bad news can distract you; sickness or premenstrual tension* can make you unreasonable; the presence of other people can injure your objectivity; an unfriendly environment can make you nervous; if you are hurried (maybe it's the wrong time) you may negotiate with impatience. Therefore, before negotiating, make certain that the time, the place, and the circumstances will be optimal. If they are not, delay the negotiations for a few days.

---

*See Part II for a discussion of premenstrual tension.

Assignment 32

1. Begin your negotiations to reach a mutually satisfying allocation of power and authority. *Do not hurry. Take as many days as necessary. A majority of couples need two or three days. Some require four or five.*

2. Continue carrying out your Performance Pact.

3. Continue Intimate Time.

Assignment 33

Yesterday you agreed on a satisfying allocation of authority and power in your relationship. Earlier in the course of instruction you negotiated and agreed on which performance behaviors each spouse is to do for the other; and before that you negotiated and agreed upon which cherishing behaviors you wanted to exchange.

If you have accomplished all the above, if you also manifest the "as if" attitude (the Law of Positive Expectations), and if you now are using the various communication avenues required in a clear-eyed, reciprocal relationship— then it is probable that the level of satisfaction of your marriage has become elevated and that your sense of equality and security has been strengthened.

You may now realize and enjoy a "good relationship," as defined in Assignment No. 2. However, that is for now only! *Everything changes. The overcoat which keeps you warm in January would make you uncomfortable in August. The light clothes you wear in August are inappropriate for January. When you remove your overcoat in the spring, you are adapting to inexorable change. When you store your light summer clothes in the autumn and put on warmer clothes, you are again adapting to inexorable change.*

It is the same in marriage. Whether you like it or not, there will be changes in yourself, your spouse, the circumstances of your lives, and the dynamics of your relationship. But—the thing you want to stay constant is the high satisfaction level of your "good relationship." This means you must adapt to inevitable changes.

How does one do this?

Answer: The same way you have done it while going through this manual. At regular intervals you must define the current situation and then adapt to the changes which have occurred. You can adapt by negotiating.

But first, let us examine the changes which will affect your lives.

THE ASSIGNMENTS

1. Study "The Changes to Which Spouses Must Adapt, as Individuals and as a Couple."

2. Study "A Schedule for the Future."

The Changes to Which Spouses
Must Adapt, as Individuals and as a Couple

Almost all married people say, "Of course, it's obvious that we will have to adjust." However, few spouses realize that their mutual and individual survival depends on their ability and willingness to adapt; that they will have to adapt to frequent and dramatic changes on a month-to-month, year-to-year basis.

When John Average-American of 1971 married Mary Average-American, he was 23.1 years old and she was 20.9. If they manage to get through life without a divorce, and if they hold to the statistical average of that year, they will bear "two and a half" children. Their marriage will terminate upon John's death at 67.1 years, which gives Mary some 5.3 years of widowhood should she live to the final day of her statistically average 72.4 year lifetime.*

This means that on their wedding day, Mary and John Average-American can look forward to 44.0 years of married life together.

During these years Mary and John will experience, and often be shocked by, (1) changes in themselves; (2) unexpected upheavals in the world and in society; (3) and unforseen changes in their relationship with friends and family, and (4) unanticipated changes in their circumstances. Examples of these inescapable and uncontrollable conditions include old age, the debacles resulting from a war or a depression, and the apparently strange attitudes of the younger generation.

On their wedding day neither Mary nor John have the slightest realization of the inexorable changes which will challenge them for the rest of their lives. However, changes will come. If the marriage is successful, Mary and John will adapt themselves to the social, cultural, economic, technological inovations, and marital changes which challenge all people. If they do not adapt (compensate), their "systems" will be out of balance.

*Bureau of the Census, *Statistical Abstract of the United States* (Washington, D.C.: U.S. Dept. of Commerce, 1973), p. 63, 69.

If they desire to adapt to changes, the spouses first must be aware of them. Therefore, let us examine a few of the inescapable changes:

If John and Mary follow the statistical average, one or both of them will change jobs from three to six times during her/his lifetime.

Mary and John, as an average couple, will move their residence an average of fourteen times during their lifetimes (one out of every five Americans changes addresses every year). This will require Mary and John to adapt to new living quarters, to find new friends, doctors, dentists, babysitters, all-night grocery stores, etc.* Their children will be uprooted and will suffer the abrasion of entering new schools and having to make new acquaintances, which will have an impact on the home environment and, along with other changes, tend to throw the family "system" out of balance.

However, even if John and Mary stayed in one residence and with one job, they still will have stressful changes imposed upon them. Let us observe the possible sequence of events:

At the beginning, their relationship probably will not be private. During the days before their wedding, their families will try to supervise much of their lives, as rehearsal and guidance for their married lives. Their parents' inclinations to traditional ritual are likely to be effective: seven out of eight American marriages take place in churches or synagogues; and the climate of the rituals encourages "romance" in the fairytale tradition.

However, after the wedding, when Mary and John are living in the privacy of their own home, the romance and "oneness" which were the characteristics of their courtship will be changed. Different behaviors will take their place.

Soon after the marriage, the "romantic" behaviors of the courtship will subside. John and Mary will experience aggressive interactions. Each will want the marriage to fulfill his or her own unique expectations. Each will try to manipulate and control the other.

*Vance Packard, *A Nation of Strangers* (New York: David McKay, 1972), pp. 6, 8.

Almost daily confrontations will cut away at their previously romantic relationship. Everyone and everything seems to get into the acts of discord. How much will John's mother have to say in their relationship? Will Mary still want to go out for a beer after her bowling team meets on Tuesday nights? Will John really expect Mary to do the ironing while he watches Monday-night football?

If John and Mary can adapt to the enormous changes which come when the fairy tales of courtship are eliminated by the realities of marriage, they may have a few peaceful years in which to enjoy one another before new changes come. However, soon they probably will be visited by an event for which they are unprepared: A child will be born.

The happy parents of the newborn will suffer disillusions about parenthood before their baby's first lung-clearing cry is heard. J. Richard Udry puts it this way:

> "Couples are so ill-informed on the nature of babies and children and so ill-prepared for their impact on the marriage relationship that they often experience a crisis with the arrival of the first child. The child transforms the marital interaction, interferes with marital intimacy, and complicates life far beyond what misty-eyed parents-to-be expect.*

Before the baby's appearance, John and Mary went where they wanted to whenever they felt like it. Their mobility and experiences were limited only by their jobs and their finances. Both probably worked and had enough money to splurge independently. Even while getting pleasure from these independent adventures, each probably enjoyed the satisfaction of being the center of the other's attention. However, the arrival of the baby changes their relationship dramatically and *irreversibly*. At first the new circumstances probably affected Mary more than they did John. As found in the research of Helen Z. Lopata,

> the event causing the greatest discontinuity of personality in American middle-class women is the birth of the first child, particularly if it is not immediately followed by a return to full-time

The Social Context of Marriage (Philadelphia: J. B. Lippincott, 1971), p. 439.

involvement outside the home. It is not just a 'crisis' which is resolved by a return to previous roles and relations, but an event marking a complete change in life approach.*

For the woman, motherhood means confinement to a small space and a narrow band of tasks, increasing economic dependence, the addition of responsibility and strain. *Along with these burdens the woman has expectations of rich compensations: a new maturity, status, and respect.* She is frustrated when these expectations are not fully realized.

For the man, fatherhood means an increase in responsibility both at home and at work. Simultaneously, he feels almost like a stranger in his own home. His wife doesn't seem to be as concerned about him as she used to be. Much of her affection and interest is now withdrawn from the husband and directed to the child.

The advent of children usually changes the interactions between the spouses and results in reduced marital satisfactions.

Some years go by: When the children grow old enough to look after themselves a bit, the wife is likely to go back to part-time employment. The lessening of the strain of full-time child-rearing, together with a return to more varied human contacts, will probably increase the wife's enjoyment of herself and her husband. However, during the years of child-rearing, because she had to do many new things alone, she developed an area of independence which is now applied to activities outside the family; now she will insist on her husband's acceptance of these changes. To keep the relationship in balance, he must adjust to his changed wife: a woman who is not always home anymore, who is not always available to do the many things to which he may have become accustomed; and who even occasionally announces that she will be flying here or there for a sales meeting sponsored by her firm; or that she will spend a weekend in a retreat run by the local women's club.

Five to ten more years go by. Our typical couple, Mary and John, settle down to the middle years of their marriage. Soon they will experience another shock, as the children leave home and start their own independent lives. (Men and women who

*Occupation: Housewife (London: Oxford University Press, 1971), pp. 200–201.

marry today have, respectively, eighty-eight and ninety-four chances in a hundred of living to see their children marry and leave home.) At about this time, Mary and John must also cope with the approaching syndrome of senior citizenship. Their jobs are beginning to diminish in importance; the oncoming of old age reduces their physical vigor; and their outside-the-home activities are less hectic.

In their senior years, John and Mary again begin to focus strongly on their relationship in a new way. If they have tended to drift apart over the preceeding fifteen to twenty years, the renewed (often forced) intimacy may make them uncomfortable. As Morton Hunt notes, they may have "simply been too busy to notice that they were no longer friends, until their aloneness made it obvious."* With still more than twenty years of life together before them, they may have to *start* the process of relationship-building, almost as they did on the day of their marriage. To make this clear, let us review in detail their lives together.

During the courtship and the early years of marriage, if John were as traditional as we have made him out to be, preferred to have Mary await his decisions in a great many areas. Ten years later, when their dishwasher breaks down for the third time, he will expect Mary to decide for herself whether or not to go ahead with a thirty-dollar repair job. Three dishwashers later, when John has much less with which to occupy himself, he may feel offended if Mary does not leave the actual repairing to him.

Mary might have wanted to do all the cooking herself during the first year or two of the marriage, regarding John's ventures into her culinary domain as encroachment or as indirect criticism. Later, when she has been exhausted by the demands of her growing family, she may consider it a blessing when John takes over half the cooking. But in middle age, when her interest in activities outside the home is less, she may wish to have the kitchen to herself again. There may be many such reversals because by this time many of Mary's basic values probably have changed, whereas John's may not have.

Her Infinite Variety (New York: Harper and Row, 1962), p. 219.

There always will be adjustments to be made. Fortunately, adaptability is something which can be learned by almost any two people who desire a successful marriage. Adaptability involves identifying the changes as they come and keeping the marital system in balance by adapting to the changes. This means negotiating and making new agreements (contracts) at regular intervals, perhaps several times a year. You know how to do this. What you need is to make up a "good relationship schedule," mark it on your calendar, and then carry it out.

A Schedule for the Future

CHERISHING BEHAVIORS

We suggest that within the next week you make up new lists of Cherishing Behaviors which you would like to exchange. (See Assignments 5 and 6 for an efficient method.) You may want to use your original lists, with only a few additions. That's fine. Rewrite the lists and keep them in your notebook. However, there is no need to keep records from now on. We suggest that you revise your cherishing behaviors lists every six months.

PERFORMANCE BEHAVIORS

It has been several weeks since you agreed on your performance pacts. It is advisable to renegotiate your performance pacts during the next week or so (see Assignments 18, 19, 20, and 21). Do this every three months or so. It is probable that you will want to renegotiate only one or two performance behaviors. There is no need to keep records as long as your new pacts are written in your notebook.

AUTHORITY AND POWER ALLOCATION

You have just completed this in Assignment 32. in which you have agreed on *when* each of the sensitive activities would be renegotiated. In the case of Mary and John, the renegotiating dates varied. Yours probably will also vary. However, we suggest that six months from now you renegotiate the entire group of authority and power behaviors—the full list of items—and that in the future you renegotiate once a year.

INTIMATE TIME

We suggest that this be done at least once a week from now on. Many couples enjoy it more often. (See Assignment 1.)

KINESICS (BODY LANGUAGE)

Make an effort to increase your vocabulary of kinesic behaviors. Once a month the entire family might play "charades" on this subject. Go through the known repertoire, with each family member adding others s/he has observed in others or in her/himself.

PHYSICAL HEALTH

See Part II, which follows.

Part II

THE PHYSIOLOGICAL

FACTORS

The following is based on material recently pre-sented by William Lederer in a series of lectures at the International College of Applied Nutrition, the International Academy of Preventive Medicine, and the European Academy of Preventive Medicine (Paris).

A Physical Condition Which Can Provoke Negative Behaviors and Thus Contribute to Marital Discord

Clinical evidence suggests that a significant proportion of unhappy couples suffer from subtle physical pathologies *of which they are not aware or which are not considered very important.* However, these conditions can cause one or both partners to experience chronic fatigue, irritability, or even unpredictable, irrational behaviors, and thus may bring about negative behaviors in the relationship.

There are, of course, many physical ailments which can negatively influence human behavior. Some of these are obvious—those involving pain or a deterioration of tissues, or such conditions as alcoholism or drug addiction. But it is not the function of this book to catalogue all of the known ailments which influence human behavior.

However, it is the purpose here to discuss, to call attention to, and to provide help in correcting, a specific pathological condition which may be present in one or both partners in many unhappy relationships; a pathological condition which frequently provokes discord and which is often overlooked and neglected.

Fortunately, diagnosis and cure (or control) is often relatively painless, inexpensive, and quick; and when this has been accomplished, psychological therapy (which usually is still needed) becomes more effective and rapid. However, if the ailment is not cured or controlled, the effectiveness of psychological therapy can be greatly diminished, and in some instances it may be made useless.

In 1963, Dr. Don D. Jackson and I (in seeking data for our book *The Mirages of Marriage**) completed four and a half years of research on 278 young and middle-aged married couples.

We observed that couples with troubled relationships had a far higher incidence of a fatigue-irritability-irrational behavior syndrome than did couples with satisfying marriages.

The symptoms, some of which were present in more than half of the troubled marriages, included the following (of course, very few of the people we studied displayed *all* of them):

> fatigue
> depression
> unpredictable temper tantrums
> moodiness
> forgetfulness
> insomnia
> headaches
> anxiety
> irritability
> crying spells
> indigestion
> compulsive eating
> sensitivity to noise and light
> pain in muscles and back
> difficulty in concentrating

*Dr. Jackson and I wrote *The Mirages of Marriage* New York: W. W. Norton, 1968) as an effort to redefine the marriage process realistically.

The symptoms usually were sub-clinical, that is, not readily identifiable by commonly used diagnostic tests and diagnostic observations. The symptoms, often vague and seemingly unrelated, did not suggest a well-defined pathology according to established and traditional medical definition. The indicants were chronic but periodic. The seemingly unrelated discomforts seldom were sufficiently acute to cause the spouses to pursue the problem aggressively with a physician. Even though the discomforts generally were neglected, they often appeared to provoke depression, irritability, and other negative behaviors which were abrasive to the marriage. Most of those people who went to physicians reported that their doctors did not give a specific diagnosis or a specific treatment. The physicians frequently advised them to take a vacation, ingest tranquilizers, or see a psychiatrist.

Although our 1963 research produced convincing data on the prevalence of the fatigue-irritability-irrational behavior syndrome in bad marriages, we did not pursue the problem further. We did not, at that time, perform diagnostic physical examinations on the unhappy persons to determine the etiology and possible therapy or control.

However, since then, extensive research and clinical work by others has confirmed that in many unsatisfactory marriages one or both partners exhibit some of the symptoms mentioned above, and that their incidence is higher in unhappy than in happy marriages. Further, when the physical *causes* of such symptoms are eliminated, psychological therapy usually becomes more effective; in some cases little or no psychological therapy is needed.

Scientists who performed etiological research observed that many of the people who exhibit the fatigue-irritability-irrational behavior syndrome often have abnormal levels of toxic metals—such as lead, arsenic, mercury, cadmium, and aluminum—in their bodies, causing a biochemical imbalance.

Scientists also discovered that the syndrome can be provoked by people eating, touching, or breathing allergens to which they are sensitive. The allergens can be of a hundred different varieties; some of the most frequent are milk, sugar, coffee, chocolate, house dust, chemicals, wheat, corn (even in the form of envelope glue or corn oil), eggs, yeast, alcohol, citrus fruits, peanuts, tomatoes, chicken, onions, potatoes, to mention only a few.

Other etiological factors contributing to the syndrome are depressed or elevated amounts of minerals or vitamins in the body, and general malnutrition.

Any of these conditions may cause a biochemical imbalance in the body, resulting in behaviors which mimic those seen in schizophrenia.

Such imbalances cannot be blamed entirely on food. In this technological culture of the late twentieth century, we are exposed to an environment for which our physical systems have not yet developed full defenses. Harmful substances enter our bodies not only in the foods we eat but also from the invisible contaminations in the air we breathe, the water we drink, and the substances we touch. The ingesting or absorbing of many of these can throw us out of biochemical balance and can provoke a state of impaired functional ability. *Dysfunctions frequently are exhibited via negative behaviors and unstable emotional patterns before the ailments are signaled by physical pain.*

When negative behaviors and unstable emotional patterns occur, the partners in a marriage have extra difficulty coping with the problems and natural stresses to be expected in all intimate, long-range relationships.

Of course, biochemical imbalance can bring about dysfunctions other than behavioral and emotional. According to Marshall Mandell, M.D., a clinical ecologist, the presence of allergens in the human body can bring about colitis, multiple sclerosis, migraine headaches, epilepsy, heart disease, pulmonary embolus, ulcers, emphysema, asthma, schizophrenia and other conditions. Dr. Mandell's data come from over three thousand documented cases.*

An associate of Dr. Mandell, Dr. Anthony Conte, discussed the tests he made in collaboration with Dr. Mandell, concerning arthritis, "Using rigorous, scientific protocol, and the first dou-

*At the Allen Mandell Biologic Medical Clinic, Norwalk, Connecticut (telephone: 203-838-4706).

ble-blind study, we have established beyond doubt that arthritis
is an allergy-related disease which can be treated by eliminat-
ing certain foods from the diet and by controlling factors such
as house dust."

"The papers concerning these tests were read to the American College of Allergists in 1979," Dr. Mandell told the author.
"Also, abstracts of the papers were published in the January
1980 issue of *Annals of Allergy* as well as in *Medical World
News* and other publications. We found in our double-blind test
that arthritis is one part of a multifaceted allergic disorder.
Symptoms of fatigue, migraine, asthma, colitus are common,
and I find it tragic that patients go from one specialist to another
seeking treatment for what they believe are symptoms of different ailments, when in 85 percent of the cases treating arthritis
as an allergic disorder would be effective for all or most of the
symptoms. When the allergens are identified and eliminated,
about half of the arthritics require no further medication."

Dr. Mandell's findings are recent and not yet well known to
the medical profession at large. His is a revolutionary treatment
which, unlike some, can do no harm if unsuccessful.

An increasing number of physical ailments which provoke
negative behaviors are being associated with biochemical imbalance. On July 5, 1980, United Press International reported as
follows:

> LONDON (UPI)—Many migraine headaches might be caused by
> food allergies, *The Lancet,* the British weekly medical journal,
> said Friday.*
>
> *The Lancet* published a research report on a two-year test car-
> ried out at two London Hospitals, the National Hospital for Ner-
> vous Diseases and the Middlesex Hospital.
>
> The tests showed as many as two-thirds of migraine attacks
> might be caused by foods.
>
> During the two-year test a group of 47 patients with migraines
> were put on special diets in rotation. Each diet omitted a single
> food from a list including milk, cheese, eggs, chocolate, shellfish,
> fish, oranges, rice and apples.
>
> Thirty-three patients completed the program and in 23 cases

*The Lancet, 5 July 1980.

the migraine attacks were traced to one or more foods, the report said.

Blood tests showed the patients had antibodies to the foods concerned and confirmed the allergies, the report said.

It said total elimination of the guilty foods from the diets of the 23 caused relief, mostly complete, of the migraine symptoms within two weeks.

Many physicians, including Arthur L. Kaslow, M.D., of the Kaslow Medical Self-Care Centers in California;* William H. Philpott, M.D., of the Ecology House Clinic, Oklahoma City; and Theron Randolph, M.D., Chairman of the Department of Clinical Ecology at American International Hospital in Chicago, have made similar findings about a wide variety of diseases in which symptoms sometimes provoke negative behaviors as well as physical disabilities.

However, in this book we are concerned only with the negative behaviors and irrational behaviors which can be harmful to a marriage (or any long-range relationship). Admittedly, there still is controversy on the subject. This is to be expected. For centuries, people who behaved irrationally were diagnosed frequently as mentally ill. There was little awareness by the medical profession of the fact that irrational behavior provoked by physical deficiencies or ailments. We now know that irrational behavior is not necessarily caused by mental disease or past psychological conditioning. It can be—and frequently is—a symptom of malnutrition or of the presence of allergens or other disrupting substances that the body is attempting to reject. For example, the deficiency disease pellagra is characterized by hallucinations and behavior resembling schizophrenia. Until 1925, advanced cases were sent to insane asylums to be treated by psychiatrists. There were few cures. In some parts of the South, the disease was common enough to be considered endemic.

In 1925, Dr. Joseph Goldberger of the U.S. Public Health Service discovered that pellagra resulted from nutritional deficiency (biochemical imbalance). The patient simply was defi-

---

*His book *Freedom from Chronic Disease* discusses the variety of ailments which can result from biochemical imbalance—as well as their cure. (New York: St. Martin's Press, 1980.

cient in one common vitamin, B3 (niacin). Pellagra cannot be cured by psychiatric treatment or by drugs. But it is cured easily and quickly by eating foods rich in Vitamin B3. One can imagine the effects of the schizophrenia-like pellagra on a marriage.

Scurvy and rickets also are ailments whose symptoms include behavioral aberrations. Both can be cured easily by dietary means.

Of the couples who have sought marital assistance through the Behavior Research Institute, over half suffered from some sort of biochemical imbalance. Similar clinical observations have been made by physicians and other practitioners in all parts of the United States.

Dr. David Hawkins, medical director of the North Nassau Mental Health Center, in Manhasset, New York, says: "About fifty percent of the couples who apply to us for marital counseling have serious nutritional problems. Sometimes malnutrition is the total cause of their marriage problems. At other times, nutritional deficiencies aggravate existing conditions. All couples were amazed at the difference a simple change in diet made in their lives. They went from being testy to being amiable."

Dr. Mary Jane Hungerford, director of the Santa Barbara branch of the American Institute of Family Relations, says: "Nutrition is involved in ninety percent of my cases. In seventy-five percent it is a major factor. Almost all of my patients complain of fatigue—one of the first signs of poor nutrition. It certainly is the basis of many fights."

Dr. Michael Lesser, psychiatrist and author of *Nutrition and Vitamin Therapy,* said that in his experience psychotherapy can be shortened by half if attention is paid to nutritional deficiencies. When the deficiencies are corrected, the resulting proper nutrition does what tranquilizers are often used for: making the patient more amenable to psychotherapy.*

It is important to understand that "malnutrition" means more than "not enough food." It *can* mean that, but more often it is a matter not of the quantity but of the nature of what we

---

*We recommend Dr. Lesser's book, *Nutrition and Vitamin Therapy* (New York: Grove Press, 1980).

ingest. It includes the taking into the body of allergens, toxic metals, fumes, improper levels of vitamins and trace minerals —in short, a condition of biochemical imbalance. This subtle debilitation can affect people of all ages and social groups.

What does all this suggest?

It suggests that *anyone* who has unexplained physical or emotional discomforts should take steps to find out whether s/he is suffering from biochemical imbalance. For readers of this book, however, the "anyone" refers specifically to couples who want to improve their relationship.

Several examples of unhappy couples who suffered from biochemical unbalance follow.

Example 1: A young couple in Vermont had been married almost two years. They were "back to nature" enthusiasts, and for two years they had lived in a log cabin which had no electricity. They heated the house with wood and cooked on a wood stove. Their relationship was a joyful one. The husband worked as a tree surgeon and looked after the livestock at home, as well as a big garden. The wife had a passion for cooking, and did so with love. Her baked goods and preserves won blue ribbons at country fairs.

When this couple decided to have children, they moved into the village. Their new home had all modern conveniences. Shortly after moving, their troubles began. The wife complained she "was sick and tired of cooking." She began preparing "fast" meals only. She became irritable, sometimes hysterical, and suffered from occasional migraine headaches. She began smoking again (she had stopped two years earlier because the cigarette smoke caused her husband to suffer from a respiratory ailment). He now angrily nagged her to stop smoking. Within weeks their quarreling had become fairly constant. They believed that perhaps they had made a mistake in getting married.

The wife saw their family physician. He examined her, found nothing physically wrong, and suggested that the problems might be caused by the stress of moving and getting ready to raise a family. Perhaps seeing a psychiatrist might help.

The wife consulted a young woman doctor, who believed that the sudden change in the wife's behavior suggested an allergen.

Upon testing the wife it was learned that the wife was highly allergic to the cooking gas in their modern stove.

The gas stove was replaced it with an electric one. Within days the wife's symptoms were gone, and the couple's good relationship resumed.

Example 2: Another couple had had four years of a highly satisfactory marriage. Then the husband received a big promotion into a new department in his company with which he was not familiar. Everyone was watching to see how he would do, and there were rumors that he was being groomed for a big executive position. The situation placed him under enormous and constant stress. To keep his energy high he began drinking coffee and eating cookies and doughnuts frequently throughout the day. But instead of being sustained by his coffee and snacks, he became more fatigued and tense. So he increased the amount of coffee he drank and the sweets he ate. He became depressed and irritable, and even more fatigued.

At home he was now sullen and nagging. In subtle ways he blamed his wife for the condition he found himself in. The more helpful and understanding she tried to be, the more quickly he blew up. Finally his wife requested that they both have complete physical and psychological examinations.

A series of diagnostic tests indicated that the husband was unusually sensitive to sugar, coffee, and the highly processed flour used in doughnuts, cookies, and most commercial breads. What he had been eating and drinking in such enormous quantities had caused his blood sugar level to drop. Low blood sugar levels can bring about depression, fatigue, and many other seemingly unrelated discomforts. Paradoxically, when people are sensitive to sugar, eating large quantities of it will reduce the body's sugar level to a point which often causes negative symptoms.

The husband stopped his excessive coffee drinking and snacking on sweets (substituting other things to which he was not sensitive) and increased his energy levels through exercise and nutrients prescribed by his clinician. He became better able to handle his job, and his satisfaction with his home-life returned.

Example 3: A husband was chronically fatigued and irritable, and sometimes complained of numbness in his fingers during cold weather. He frequently had severe halitosis, headaches, and insomnia. His wife began to hate the thought of her husband coming home, dreaded the scenes at dinner, and found that their once satisfying sex experiences now were unpleasant. A divorce seemed inevitable.

The physician to whom the husband went found no specific physical pathologies. He believed that the husband's physical symptoms might be the result of the stress of a bad marriage, and he suggested psychiatric help, along with marriage counseling.

The counselor to whom the couple went belonged to a clinic, and both spouses were given a complete diagnostic work-up (we will tell you about these tests later). The work-up showed that the husband was deficient in thiamin (Vitamin B1) and niacin (Vitamin B3), and had a depressed level of phosphorus. *The husband was in a bad state of biochemical imbalance—the ailment which is common to all the examples we are presenting.*

The doctor at the clinic supplemented the husband's diet with appropriate nutrients and additional proteins. He recommended that the husband reduce his intake of alcohol, coffee, and sugar, and he prescribed a moderate exercise routine. Within two months the husband had regained his biochemical balance and health. After a short period of behavior-modification therapy (to help the couple eliminate their acquired abrasive behaviors), the marriage improved.

Example 4: A couple had been married for twenty-one years, and both considered it a satisfying relationship. When the children were able to look after themselves, the wife got a job in a factory.

A few years later the wife gradually seemed to lose her zest for life. She was fatigued, listless, and had frequent abdominal discomforts. Food did not interest her, she lost weight, and no amount of luxury eating over weekends helped. Her gums became inflamed. She became irritated easily and began throwing temper tantrums at home. The discord in the relationship increased.

A routine physical examination at the factory's clinic did not produce a specific diagnosis. It was suggested that the wife was suffering from menopause, aging, and stress in general (one of her sons had been killed in Vietnam), and that perhaps a six-month leave of absence and a paid vacation might help her.

When her gums became so inflamed that eating was difficult, the wife went to a dentist. He supervised a diagnostic work-up. From an analysis made of the wife's hair it was learned that her body contained high levels of the toxic metals lead and aluminum, both of which she was exposed to in her work. She was treated, and the levels of aluminum and lead were reduced, and she was put on a therapeutic diet. In five months her physical and emotional condition were greatly improved, and so was her marriage.

In summary, then, without the partners being aware of it, a biochemical imbalance can bring about a negative, irrational behavioral exchanges. It can blur an individual's perception—especially concerning the quality of her/his *own* and her/his partner's behaviors and motivations. This in turn can impede improvement in the marriage regardless of the spouse's good intentions and strenuous effort, and it can diminish the effectiveness of skilled psychological therapists who are not aware of their client's physical condition.

To assist you in your marriage, we have prepared a simple test. The test is a rapid way of *estimating* whether or not your behavior and emotions are being influenced by a general, subtle physical pathology resulting from a biochemical imbalance. We emphasize that the test is only a general indicator, a means of alerting you to a *possible* condition.

The Test

This is a test which can help you make an intelligent guess as to whether or not you suffer from biochemical imbalance which may be caused by malnutrition, a physical condition which may be having a destructive effect upon your marriage.

The test was developed after many distressed couples had been observed. There are half a dozen similar tests currently in use by clinicians.

Instructions

On the following pages are 53 questions. Answer them in the following way:

If the answer to a question is "never," put the number 0 in the "Never" column.

If the answer to a question is "hardly ever" (about once a month), put the number 1 in the "Hardly Ever" column.

If the answer to a question is "moderately" (about once a week), put the number 2 in the "Moderately" column.

If the answer to a question is "a lot" (more than once a week), put the number 3 in the "A Lot" column.

	Never	Hardly Ever	Moder-ately	A Lot
Do you feel tired an hour or two after meals?				
Are you inclined to find fault with many of the things your spouse says or does?				
Do you have headaches?				
Do you have cravings for sweets or salt?				
Do you feel exhausted and have a hard time getting up after eight or more hours of sleep?				
Do you have muscle aches or backaches?				
Do you have cravings for coffee?				
Do you have trouble falling asleep?				
Do you have nervous fatigue?				
Does your husband believe you are excessively moody and irrational during your premenstrual period?				
Does your wife say you are unbearable during your periodic "bad" days?				
Do you sometimes faint or feel like fainting?				
Do you lose your temper unexpectedly?				
Do you have constipation?				
Do you experience impotence?				

Column annotations above "Hardly Ever", "Moder-ately", and "A Lot": *Once a Month*, *Once a Week*, *More Than Once a Week*

	Never	Hardly Ever	Moderately	A Lot
Do you have crying spells?				
Do you blame others when things go wrong?				
Are you unusually sensitive to noise or light?				
Do you have sudden spells of intense hostility to your spouse?				
Do you withdraw from your family?				
Are you obsessed by the idea that things must be done your way in the house?				
Do you have a strong urge to eat when you are nervous?				
Do you have loss of appetite?				
Do you have temper tantrums?				
Do you feel especially good after having breakfast?				
Do your gums bleed?				
Do you have to have coffee in the morning before you feel you can start moving?				
Are you restless, unable to sit still and do nothing?				
Are you absent-minded?				
Do you get depressed?				
Do you have cold hands or feet?				
Do you have difficulty making decisions?				
Do you become suddenly irritable?				
Do you have dizzy spells?				

	Never	Hardly Ever	Moder- ately	A Lot
Do you forget things?				
Do you get frantic over problems which your spouse considers routine?				
Do you have cravings for alcohol?				
Do you get winded after little or moderate exercise?				
Are you anxious or afraid over routine problems?				
Do you feel hungry between meals and have a compulsion to eat?				
Do you at unexpected times feel physically weak?				
Do you exhibit hostile or anti-social behavior, believing that others are incompetent or immoral?				
Do you feel in an elevated mood immediately after eating?				
Does your spouse suggest that you see a psychiatrist or other medical practitioner because of your irritability or moodiness?				
Do you have weak spells?				
Do you have indigestion?				
Do you get impatient?				
Do you feel confused?				
When you miss a meal or the meal is late, do you get tense and cross?				
Do you scold if anyone messes up the house even a little?				
Are you nervous or irritable when you are hungry?				
Do you wake up at night and find that you cannot get back to sleep?				
Does eating sometimes help you get back to sleep?				

Instructions for Final Scoring

Add all the numbers in all the columns.

If your score is 70 or over, have your counselor or doctor arrange a complete physical examination for *both* of you. It is important that both partners get the diagnostic work-up, even if it appears that only one might need it.

Do not be alarmed if your score is 70 or over. If you have such a score and happen to be in biochemical imbalance, you are not unique. As mentioned earlier, many clinicians have reported that over half of their clients who seek marital counseling have unrecognized biochemical imbalances which contribute to the discord in their marriages.

What to Do

If a routine physical check-up does not result in a specific physical diagnosis, we urge you to request your doctor or counselor to arrange for the following diagnostic tests:

1. Allergy Tests

There are many new advances in the field of allergy detection and cure. For example, some allergists and clinical ecologists can identify many allergies simply by putting extracts of the various probable allergens under the tongue and observing the patient's reaction.

Testing for the more common allergies can be done by you at home. One method is Dr. Coca's Pulse Test or variations of it. You will find complete instructions for the Coca Pulse Test on page 229. The instructions are condensed from Dr. Coca's book.*

Another way of identifying allergies at home is described by Marshall Mandell, M.D., in his book *Dr. Mandell's 5-Day Allergy Relief System.*†

---

*Arthur F. Coca, M.D., *The Pulse Test—Easy Allergy Detection* (New York: Arco, 1978).

†*Dr. Mandell's 5-Day Allergy Relief System* (New York: Thomas Y. Crowell, 1979).

2. A Hair Analysis

This is a fairly new diagnostic tool which, after early technical difficulties, has in the last five years become an accurate means of measuring the amounts of toxic metals (lead, arsenic, mercury, cadmium, and aluminum) in the human body. The hair analysis also measures the levels of nutrient minerals (calcium, magnesium, sodium, potassium, copper, zinc, iron, manganese, chromium, cobalt, lithium, molybdenum, nickel, phosphorus, selenium, and silicon). Certain factors, of course, can influence the readings—the client's hair coloration, age, body size, sex, hair dyes. Laboratories now are compensating for these factors, but even if there are variations in the values measured (and there sometimes are), the general ratios between values indicate possible trends which aid medical clinicians in reaching a diagnosis. Appendix II gives further information on hair analysis, along with a list of laboratories to which your clinician can mail the hair sample.

3. A Full Blood Profile

This is a more complex and thorough analysis than the standard blood test. Your doctor will be able to arrange for you to have the profile made.

4. A Five- or Six-Hour Glucose Tolerance Test*

The glucose tolerance test is used to determine the presence of a low blood sugar condition (sometimes called hypoglycemia). More information is given in Appendix II, including a self-observation chart. Your description of how you feel and behave during the glucose tolerance test is just as important as the measurement of the blood sugar levels, and sometimes more so. The self-observation chart should be given to your doctor, along with the laboratory report.

At this writing, the laboratory work-up we suggest will cost about $150. The practitioner may charge another $50 for interpreting the tests and prescribing treatment.†

Your doctor or therapist may want to refer you to someone

---

*This may not be needed if information gained from other tests leads to the elimination of symptoms.

†Based on 1980 charges for a full work-up at the laboratory of the North Nassau Mental Health Clinic, Manhasset, New York.

specializing in the diagnostic skills which have been effective in identifying various forms of biochemical imbalance. Your doctor, your therapist, or you can get a list of practitioners in your vicinity by writing or telephoning:

United States
The International Academy of
  Preventive Medicine
10409 Town and Country Way,
Suite 200
Houston, Texas 77024

The Orthomolecular Medical
  Society
2340 Parker Street
Berkeley, California 94704

The Academy of Orthomolecular
  Psychiatry
1691 Northern Boulevard
Manhasset, New York 11030

The International College of
  Applied Nutrition
Box 386
Le Habra, California 90631

The Huxley Institute
1114 First Avenue
New York, New York 10021

Del Stigler, M.D.
Secretary of the Society for
  Clinical Ecology
Midtown Medical Building
2005 Franklin Street, Suite 490
Denver, Colorado

International Academy of
  Metabology, Inc.
2235 Castello Street
Santa Barbara, California 93105

American Academy of Medical
  Preventics
2811 L Street
Sacramento, California 95816

Biomedical Synergestics
  Institute, Inc.
434 North Oliver Street
Wichita, Kansas

Australia
Contract Allergy Services
2 Vielers Street
Paramatta, N.S.W.
Sidney
AUSTRALIA

England
Mrs. G. Henmings, Secretary
Schizophrenia Association of
  Great Britain
Llanfair Hall
Caernarvon
ENGLAND

France
European Academy of
  Preventive Medicine
2, Avenue Talma
92 500 Rueil-Malmaison
FRANCE

Canada
Canadian Schizophrenia
  Association
2229 Broad Street
Regina, Saskatchewan
CANADA S4P1Y7

Note: We cannot vouch for the skill of every clinician or laboratory listed above.

Relief of Menstrual Discomfort with Nutritional Supplements

The discomfort and negative behavioral changes which frequently accompany the menstrual process make many women feel that they have been discriminated against by nature. In addition, they are frustrated by men's inability to understand their discomfort and feeling of helplessness. Our experience indicates that in shaky marriages the incidence of quarreling is usually greater during the wife's premenstrual and menstrual period. In some case, the wife's behavior may be negative for several days. When the period is over, she frequently does not recall her behavior; this, in turn, frustrates the husband.

The physical and mental discomfort of the menstrual period is caused primarily by a temporary biochemical imbalance. During the ten days before menstruation there often is a gradual decrease in the blood calcium level. It is this which can provoke discomfort. When the discomfort comes, the stress causes the body to secrete cortisone and aldosterone, resulting in the retention of water and salt, and more discomfort.

For many women, the following nutrient program, beginning ten days before menstruation, is effective.

Take three times a day before meals:
400 milligrams of Vitamin D

250 milligrams of calcium
125 milligrams of magnesium (the calcium is retained better
with a high level of magnesium and Vitamin D)
500 milligrams Vitamin C
150 milligrams pantothenic acid (calcium pantothenate)*

If puffiness and weight persist increase the Vitamin C to 1,000
milligrams at every meal.

Well over half of the women who have used this formula
found that it either prevented or decreased premenstrual discomfort. Of those who did not benefit, many had depressed
levels of B_{12} or iron or suffered from allergies. Undoubtedly
there are other factors which we know little about.

Although results have been excellent, we suggest that you
discuss this routine with your medical practitioner, especially if
s/he has already prescribed a medication for menstrual discomfort.

Note: Males also have "bad day" cycles. To our knowledge,
there is no nutrient program that will reduce their effects. However, by careful observation a pattern of "bad days" can be
noted. Our own inadequate clinical observation shows a cycle
of twenty-three to twenty-eight days. (In some it seems to occur
every twenty-three *and* twenty-eight days.) Once a "bad day"
schedule has been established, the forewarned man must make
simply make an effort to control himself.

*All of these ingredients have been put into two capsules sold under the name
Solace 2 by Miller Pharmacal Group, Inc., P.O. Box 279, West Chicago, Ill.
60185; (312) 231-3682.

A Personal History

Soon after the writing of this manual was begun, I realized that I never had had the diagnostic tests which we had been recommending for others.

It startled me to realize this, especially as I, the so-called marriage expert, had been divorced twice. This gave me something to think about. In both my marriages, my spouse and I had worked hard to make our relationship functional and satisfying, and we had spent a large portion of our incomes paying for professional assistance. Yet during the hundred or so marital therapy sessions over the years, no therapist had given us a physiological test of any kind (or any type of test which might link discordant behaviors with one or both of our physical conditions). (The information presented above was not then generally available.)

I knew that in order to write about the diagnostic tests, I had to go through them myself, not only to satisfy my sense of integrity but also to gain understanding of the client's point of view.

So I took the tests. They cost me $140 and required some work at home (on the allergy tests).

The results of the diagnostic examinations astounded me. I learned, *for the first time,* that my biochemical condition was extremely poor. I had intermittent low blood sugar levels triggered by various allergens I had eaten, touched, or breathed.

I learned that many of the foods for which I frequently hungered (and was addicted to) would, if eaten in quantity, cause a plunge in my blood sugar level and thus impair my physical and emotional functioning. These foods included chocolate, sweets of any kind, milk, coffee, processed grains (most commercial breads, most pastas, most breakfast cereals), alcohol, nuts. Also to be avoided were certain combinations such as tomatoes and starches, and egg, mayonnaise, and wheat as combined in egg-salad sandwiches.

The hair analysis test showed that I had high levels of aluminum and lead (either of which can provoke negative feelings and behaviors) and I had insufficient levels of zinc and selenium, and that I was not retaining sufficient calcium. Such a combination (although it may not cause warning aches and pains) can result in abnormal behaviors.

As I look back at my life BDT (Before Diagnostic Tests), I can clearly see how my unpredictable (and unnecessary) depressions, fits of irascibility, fatigue, sarcastic outbursts, unreasonable demands—within my marriage and outside of it—to a large extent could have been the result of my biochemical imbalance. I now can clearly understand how it was that my desperate efforts to improve my relationships with other people failed. *True, psychological therapy also may have been needed, but by itself, in my case as in so many, psychological therapy was insufficient.*

Once I learned about my condition, I eliminated the foods, substances, and environmental conditions which had provoked many of my negative behaviors and had reduced my physical vitality. I began taking vitamins and minerals in amounts prescribed for my particular condition. I began eating mostly whole foods and fresh foods. I began exercising on a regular basis.

The change in my life was like a miracle. My health improved. At sixty-seven I felt alive and bouncy and vital. My behaviors (and my thoughts) became more positive and predictable. I became more tolerant and, I'm told, more understanding of the needs of others.

It may well be that I became marriageable for the first time.

Mind you, there may be other causes of negative behavior and physical discomforts besides biochemical imbalance, but this condition appears to be one of the most frequent causes. The process of diagnosis should begin with an examination of this aspect of the human body. Biochemical imbalance can easily and inexpensively be diagnosed, controlled, or cured.

Do not be discouraged if your medical practitioner is not immediately familiar with some of the diagnostic tests suggested here. Remember that your doctor is trained, is intelligent, and is dedicated to helping you. If you are insistent about what you want, most doctors will take steps to help you or will refer you to someone who can. The doctor will realize that fundamentally it is your responsibility to improve your health, your marriage, and the quality of your life.

Meanwhile, continue the daily assignments as scheduled in Part I. Do not delay them while waiting for an appointment with a doctor or the return of your diagnostic tests.

Diagnostic Tests

In the succeeding chapters we present information on diagnostic tests. Our purpose is to increase rapport between readers and their doctors, and to make available general information for those unfamiliar with the tests. Of course, there are other diagnostic tests, and there are many scholarly treatises in medical libraries for the practitioner who desires to research the subject.

Testing for Allergies

We suggest that the allergy test be done before the other diagnostic tests, since it may make the other tests less necessary. Both partners should take the test (and every other test), regardless of who seems to be healthy and who appears to be ailing. *It is important that both partners take all tests.*

Dr. Coca's Allergy Test

This is only one of many available tests for allergies. It is one you can do alone, but if you have a good diagnostic allergist or clinical ecologist, go to her/him first.

Arthur F. Coca, M.D., discovered a simple method for testing allergic reactions to foods and inhalants. It can be administered at home by the individual concerned.

Dr. Coca was for seventeen years the medical director of Lederle Laboratories, one of the largest pharmaceutical houses in the world. He was the founder and first editor of the *Journal of Immunology,* the foremost medical publication of its kind in the field. Dr. Coca taught at Cornell University, the University of Pennsylvania, and the Post-Graduate Medical School at Columbia University.

Dr. Coca published his findings in his book *Pulse Test* (New York: Arco, 1968; 4th printing, 1978).

Reading the entire book would be useful for those who wish to pursue the subject. However, a condensation of the methodology by David Sheinkin, M.D., Michael Schacter, M.D., and Richard Hutton appeared in their book *The Food Connection.* (Indianapolis: Bobbs-Merrill, 1978). It is reprinted here by permission of the authors.

> [Dr. Coca] learned that many symptoms affecting different parts of the body could be treated successfully as allergies. These problems included various pathologies linked to the brain, from epilepsy to emotional instability, nervousness, and depression.
>
> Four factors seemed to indicate the allergic nature of these conditions:
>
> 1. The symptoms disappeared when the implicated foods were eliminated from the diet.
>
> 2. Many patients suffered from a variety of completely different symptoms, all of which disappeared together when offending foods or chemicals were discontinued.
>
> 3. In most cases the symptoms could be brought back if the foods were reintroduced into the diet.
>
> 4. Symptoms were accompanied by an increase in the pulse rate.
>
> ### PERFORMING THE PULSE TEST AT HOME
> The pulse test is based on a simple, easy-to-prove premise: your pulse rate generally accelerates after you eat foods or inhale substances to which you are sensitive. Testing your pulse to uncover sensitivities is not difficult, but it does require a commitment on your part to a program of testing and analysis that may take a week or longer.

BEFORE YOU BEGIN THE TEST

According to Dr. Coca, the following steps must be undertaken at least two or three days before the testing begins:

1. You *must* stop smoking before and refrain from smoking during the pulse test. Tobacco is a major cause of sensitivities and can destroy the results of the test. (You can test your sensitivity to tobacco after you have finished testing foods.)

2. You must keep a written record of your pulse rate, making fourteen separate recordings a day, for at least two or three days before the test. Test your pulse at the following times: (a) just after waking up in the morning, but before getting out of bed; (b) just before each meal. (c) three times after each meal, at half-hour intervals; and (d) just before going to sleep. Each of these fourteen pulse counts should be taken while you are sitting down, except for the first one of the day; it should be taken while you are still lying in bed.

3. Remember to write down each food you eat at each meal. If you eat something like vegetable soup (which is composed of a variety of different foods), be sure to list each ingredient.

4. If you want to eat, drink, or chew gum between meals, be sure to wait at least an hour and a half after your last meal so that the snacks won't interfere with your pulse recordings. Snack foods should be treated as separate meals and listed on your chart also, along with the time you ate them. Your pulse should again be tested once before and (three times at half-hour intervals) after you eat.

Because normal pulse rates can differ significantly from individual to individual, the preparatory steps listed above are crucial. You must determine your own characteristic rates before you can actually begin testing. Your pretest recordings will be used to arrive at three different factors: (a) your lowest pulse rate each day; (b) your highest pulse rate each day; and (c) the difference each day between your lowest and highest rates (your pulse differential).

As you evaluate these pretest recordings, keep the following guidelines in mind:

1. Under normal conditions the pulse rate is fairly stable, but of course charged emotional situations and physical exercise can cause it to speed up.

2. Your lowest pulse rate during the day usually occurs before you arise in the morning. However, if you are sensitive to something in your bedroom (a feather pillow; dust in your bedding), or

even to food you ate the night before, the first rate you record may well be higher than some of the later ones.

3. If your pulse differential is greater than 12 during any single day—if, for example, your high is 82 and your low rate is 69—you may have a sensitivity to something you ate that day; if your differential is greater than 16—e.g., high of 88, low of 66—you most likely have a sensitivity; and if your pulse differential does not exceed 12 on any of the three pretest days, you are probably not sensitive to any of the foods you ate on those days.

4. If your daily high pulse varies by more than two beats from day to day, you most likely have a sensitivity. On the other hand, if your high pulse rate remains consistently within one or two beats from day to day, you have probably not eaten anything to which you are sensitive.

5. In the absence of an obvious cause (such as infection, sunburn, strenuous physical or emotional activity), a large increase over your characteristic pulse at any one recording can usually be traced to a sensitivity reaction.

THE PULSE TEST

The record you have established during the three pretest days now makes it possible to evaluate individual foods. To begin your testing, make a list of the foods you wish to evaluate. Be sure to include any food that your suspect may be causing a sensitivity reaction, as well as the foods you eat regularly. If your pretest information has indicated a probable sensitivity, use it to zero in on possible culprits. For example, if your pulse rate rose significantly after you ate a meal that included beef, potatoes, bread and broccoli, be sure to include these foods in your testing.) An increase in your pulse rate in the morning before you have eaten should lead you to suspect a chemical or inhalant sensitivity, while an increase after a specific meal tends to implicate one or more of the foods eaten at that meal.

A minimum of two days is usually required to complete the testing. To use the test effectively, follow this procedure:

1. When you wake up, record your pulse as you did during the pretest period.

2. Throughout the rest of the day, eat a small amount of a different food each hour.

3. Count your pulse just before eating and a half hour after eating each food, and record the results.

4. If your pulse rate increases after you eat a food, wait until your pulse returns to normal before testing another food. Al-

though it isn't common, it's possible for a reaction (increased pulse rate) to last several hours. At times this reaction may be erratic; for example, a higher rate may settle down, only to increase again. This pattern may recur several times before the pulse steadies. Therefore, after an increased rate, allow your pulse to normalize for at least one hour before testing the next food.

5. If you are sensitive to a certain food but have not eaten it for a few weeks, you may find no increase in pulse rate during your first test. Therefore, if you test a food you haven't eaten in a while, test it a second time two or three days later.

6. You can test chemicals and inhalants in a similar way. Expose yourself to whatever chemical you choose to test exactly as you would be exposed to it under normal circumstances. For example, if you suspect you react to the gas from your stove, stand near the stove for at least five minutes while the burners are on, and breathe normally. (It is best not to have food on the burners while you are doing this, since fumes from the food may be a source of sensitivity.) To test auto exhaust, stand outside near traffic; to test perfume, either put some on or open the bottle and inhale its fumes for a minute; to test house dust, vacuum and/or dust your house as usually; to test a pet cat or dog, hold the animal close to your nose for a few minutes and breathe normally. Record your pulse just before, just after and a half hour after the test. Note also whether any symptoms develop during or shortly after the test.

INTERPRETING THE TEST

Use the following rules as guidelines in interpreting the information accumulated from the pulse test. Note that these are not rigid rules; pulse rates and discrepancies vary from individual to individual, and exceptions do exist.

1. If a food that you eat frequently does not cause your pulse rate to increase at least six beats above your normal high pulse rate, that food is probably nonallergenic.

2. If a food causes your pulse rate to increase six or more beats above your normal high rate, it is most likely a sensitizing agent.

3. A pulse count above 84 beats per minute unaccounted for by other known factors generally indicates an allergy either to the test material or to something in the environment.

4. If your pulse differential remains high no matter what you eat, you are sensitive either to almost all the foods you have tested or to something you are constantly inhaling.

5. Pulse reactions to inhaled allergens usually do not last long and are not so severe as those to food allergens.

6. An increase in your pulse rate of less than six beats per minute during a test is usually due to an inhalant.

7. If, after you have completed your testing of foods and inhalants, you go back to smoking and decide to test it as well, note that if you are sensitive, your pulse will usually increase within fifteen minutes from the time you began to smoke.

ADVANTAGES OF THE PULSE TEST

1. Pulse testing is an accurate, objective method of uncovering sensitivities.

2. It can be used to test foods, inhalants and chemicals.

3. It is safe and free.

4. It can be performed at home.

DISADVANTAGES OF THE PULSE TEST

1. You may get withdrawal symptoms during the testing.

2. The test requires time and concentration: you must remember to make multiple recordings for at least a week.

3. For the test to be accurate, you must stop smoking.

4. You must curtail your snacking, gum chewing, etc., for the duration of the test.

5. Other external factors, such as physical or emotional activities or exposure to extraneous chemicals during the testing, may alter its accuracy.

It must be remembered that identifying an allergy does not always solve all problems. An allergy may mask other serious problems, and social pathologies may exist with and independently of sensitivity to allergens.

General Information on Allergies

Phyllis L. Saiffer, M.D., M.P.H., a clinical ecologist,* has made some valuable suggestions concerning allergies which are here reprinted, by her permission:

*Phyllis L. Saiffer, M.D., M.P.H. (practice limited to allergy), 2031 Telegraph Avenue, Berkeley, California 94705; (415) 849-3346.

WHEN TO SUSPECT THAT ALLERGY MAY BE THE CAUSE OF YOUR LACK OF WELL-BEING

1. Family history—do or did parents, grandparents, sisters, and brothers have hayfever, sinus trouble, asthma, eczema, digestive problems, or other chronic complaints which perhaps went undiagnosed as allergy? Is there alcoholism, obesity, or emotional instability in blood relatives?
2. Did you or your spouse have, as a child, colic, feeding problems, tonsil infections, recurrent earaches or ear infections, bellyaches, frequent pneumonia, colds, headaches, or learning problems? Did you have allergies as a child?
3. Do you now suffer from a collection of mild discomforts no one of which alone is enough to make you seek medical attention and which, if taken together, make you sound like a hypochondriac?

SYMPTOMS (BY SYSTEMS)

Brain

Fatigue, headache, mood swings, depression, hyperactivity, nervousness, anxiety, spaciness, excessive need for sleep, insomnia, difficulty getting started in the morning, inability to concentrate, forgetfulness, confusion. Food cravings, excessive thirst, sensitivity to odors.

Eyes, Ears, Nose, Throat

Blurred vision, light sensitivity, itchy, runny, burning, aching eyes. Sensitivity to noise, ringing in the ears, clogged or itchy ears. Runny, itchy, stuffy nose, post nasal drip, sneezing fits. Sore or itchy throat, canker sores.

Lymph System

Swollen glands, especially in the neck.

Vascular

Sweats (particularly at night), chilling, cold hands and feet, easy bruising, generalized swelling or fluid retention. Low blood pressure. Low body temperature. Labile hypertension.

Chest

Cough, wheeze, shortness of breath, racing heart, irregular heart.

Gastrointestinal

Stomachaches, nausea, gas, bloating, constipation and/or diarrhea, itchy bottom.

Genito-urinary

Frequency of urination, chronic vaginal discharge, worsening of any of the above symptoms with menses.

Neuromuscular

Muscle aching (growing pains, legaches), weakness, numbness and tingling, internal trembling, vague joint pains (palindromic rheumatism).

Skin

Itching, dryness, rashes, hives.

If the above story and symptoms, describe you, consider the possibility of an allergic etiology to your lack of well-being.

For further information for the layman, read:

Marshall Mandell, M.D. *Dr. Mandell's 5-Day Allergy Relief System* (New York: Thomas Y. Crowell, 1979).

Robert Forman, Ph.D., *How to Control Your Allergies* (New York: Larchmont Books, 1979).

Doris J. Rapp, M.D., *Allergies and the Hyperactive Child* (New York: Simon and Schuster, 1979).

Natalie Golos, *Coping with Your Allergies* (New York: Simon and Schuster, 1979).

Richard Mackarness, M.D., *Eating Dangerously* (New York: Harcourt, Brace, Jovanovich, 1976).

Instructional Courses for Physicians

Information concerning instructional courses in related techniques can be obtained from the following:

Dor W. Brown, Jr., M.D.
109 South Adams Street
Fredericksburg, Texas
78624

Russell I. Williams, M.D.
1605 East 19th Street
Cheyenne, Wyoming
82001

Basic Seminars in Clinical
Ecologic Techniques

c/o Lawrence Dickey,
M.D.
109 West Olive Street
Fort Collins, Colorado
80524

James W. Willoughby,
M.D.
Suite 1505 Traders Bank
Building
1125 Grand Avenue
Kansas City, Missouri
65106

Joseph B. Miller, M.D.
Three Office Park

273 Azalea Way
Mobile, Alabama 36609

Papers on particular aspects of these subjects are presented at meetings of the following:

The Society for Clinical
  Ecology
Del Stigler, M.D.,
  secretary
2005 Franklin Street,
  suite 490
Denver, Colorado 80205

Pan American Allergy
  Society
Dor W. Brown, Jr., M.D.
109 South Adams Street
Fredicksburg, Texas
  78624

ENVIRONMENTAL CONTROL UNITS
Dr. Theron Randolph
Zion International
  Hospital
505 North Lake Shore
  Drive
Chicago, Illinois 60611
(312) 828-9400

Dr. William Rea
Brookhaven Medical Center
8345 Walnut Hill Lane
Dallas, Texas 75231
(214) 368-4132

Hair Analysis*

Hair has the potential to become a remarkable diagnostic tool. It is easily collected without trauma on the part of the donor, it can be stored without deterioration, and its contents can be analyzed relatively easily. Trace elements, in particular, are accumulated in hair at concentrations that are generally at least ten times higher than those present in blood serum or urine and may provide a continuous record of nutritional status and exposure to heavy metal pollutants. Some drugs have already been shown to accumulate in hair, and it seems likely that other organic chemicals may be identified there when sufficiently accurate analytical techniques are developed. Hair analysis thus promises to be an ideal complement to serum and urine analysis as a diagnostic tool.†

Prior to 1970 there were criticisms concerning hair tissue analysis as a diagnostic tool, on the grounds that variables might influence the analysis. At that time the objections were valid. However, since then new methodologies and technical developments have made it possible to compensate for possible variables.

There is a large body of recent literature on the subject of hair analysis. This can be obtained most easily by writing to the hair

*This should be done through the medical practitioner, who will interpret the results and prescribe accordingly.
†"Hair: A Diagnostic Tool to Complement Blood Serum and Urine, *Science,* Vol. 202 (December 22, 1978).

analysis laboratories, which can also supply schedules of seminars and postgraduate courses on the use of hair analysis as a diagnostic tool. (A list of laboratories is given at the end of this section.)

Administering the test is simple. Remove about two tablespoons of hair from the nape of the neck, fill in the forms supplied by the laboratories, and mail. A basic analysis for about twenty-one minerals and five toxic metals costs an average of thirty-five dollars or less. The analysis is usually returned to the medical practitioner within a week. A summary of the information supplied is given in the charts below (from Doctor's Data, Inc.).

NUTRIENT MINERAL LEVEL CHART

Mineral	Deficiency Symptoms	Therapeutic Applications
Ca/calcium	heart palpitations, insomnia, muscle cramps, nervousness, arm & leg numbness, tooth decay, osteoporosis, rickets, brittle fingernails, associated with gray hair	arthritis, aging symptoms (backache, bone pain, finger tremors), foot/leg cramps, insomnia, menstrual cramps, menopause problems, nervousness, overweight, premenstrual tension, rheumatism
Cr/chromium	atherosclerosis, glucose intolerance in diabetics, disturbed amino acid metabolism	diabetes, hypoglycemia, multiple pregnancies, protein-calorie malnutrition
Co/cobalt	iron deficiency anemia	iron deficiency anemia
Cu/copper	general weakness, impaired respiration, skin sores, diarrhea in infants, copper deficiency anemia	copper deficiency anemia, baldness

Mineral	Deficiency Symptoms	Therapeutic Applications
Fe/iron	breathing difficulties, brittle nails, iron deficiency anemia (pale skin, fatigue), constipation, sore or inflammed tongue	alcoholism, anemia, colitis, menstrual problems, impaired absorption, blood loss
Li/lithium	manic-depressive disorders	manic-depression
Mg/magnesium	confusion, disorientation, easily aroused anger, nervousness, rapid pulse, tremors	alcoholism, cholesterol (high), depression, heart conditions, kidney stones, nervousness, prostate troubles, sensitivity to noise, stomach-acidity, tooth decay, overweight, protein-calorie malnutrition
Mn/manganese	ataxia (muscle coordination failure), dizziness, ear noises, loss of hearing, abnormal carbohydrate & fatty acid metabolism	allergies, asthma, diabetes, fatigue
Mo/molyb- denum	none known	copper deficiency, anemia, gout
Ni/nickel	liver cirrhosis, kidney difficulties	cancer, liver cirrhosis, kidney malfunction
P/phosphorus	appetite loss, fatigue, irregular breathing, nervous disorders, overweight, weight loss	arthritis, stunted growth in children, stress, tooth and gum disorders
K/potassium	acne, continuous thirst, dry skin, constipation, general weakness, insomnia, muscle damage, nervousness, slow irregular heartbeat, weak reflexes	acne, alcoholism, allergies, burns, colic in infants, diabetes, high blood pressure, heart disease (angina pectoris, congestive heart failure, myocardial infarction)

Mineral	Deficiency Symptoms	Therapeutic Applications
Se/selenium	mercury toxicity, pancreatic insufficiency, cardiac toxicity of drugs, aging pigment, peroxidation of fats, blood hemolytic problems, muscle wasting	cancer, mercury toxicity.
Si/silicon	bone decalcification, tendonitis, cardiovascular disease	rapid aging, abnormal skeletal formation, atherosclerosis
Na/sodium	appetite loss, intestinal gas, muscle shrinkage, vomiting, weight loss.	dehydration, fever, heat stroke
Zn/zinc	delayed sexual maturity, fatigue, loss of taste, poor appetite, prolonged wound healing, retarded growth, sterility	alcoholism, atherosclerosis, baldness, cirrhosis, diabetes, internal and external wound & injury healing, high cholesterol (eliminates deposits), infertility

TOXIC METAL LEVEL CHART

Toxic Metal	Toxicity Symptoms	Interferes with Bodily Functions
Al/aluminum	gastrointestinal irritation, colic, rickets, convulsions	irritating to gut, affects bone formation, brain convulses (high concentrations)
As/arsenic	fatigue, low vitality, listlessness, loss of pain sensation, loss of body hair, skin color changes (dark spots), gastroenteritis	metabolic inhibitor (reduces energy production efficiency), cellular & enzyme poison

Toxic Metal	Toxicity Symptoms	Interferes with Bodily Functions
Cd/cadmium	hypertension, kidney damage, loss of sense of smell, decreased appetite	heart and blood vessel structure (hypertension), kidneys, blocks appetite & smell centers, calcium metabolism
Pb/lead	weakness, listlessness, fatigue, pallor, abdominal discomfort, constipation	enzyme poison, osteoblast production, blood formation, blocks enzymes at cell level
Hg/mercury	loss of appetite & weight, severe emotional disturbances, tremors, blood changes, inflammation of gums, chewing & swallowing difficulties, loss of sense of pain, convulsions	destroys cells, blocks transport of sugars (energy at cell level), increases permeability of potassium

Important note: No matter how apparent and direct the hair analysis information may seem, it must be studied and evaluated by the diagnostician.

Even more significant than the readings of single minerals or toxic metals is their pattern. The ratio between certain minerals and metals is meaningful to the diagnostician. For example, adults having a pattern of nervousness and anxiety usually show a low magnesium/manganese ratio. These patients also often have elevated zinc and cadmium levels. When the low Mg/Mn ratio appears in children, learning disorders usually are indicated. The hair analysis chart of an individual with low blood sugar usually shows an increase in the calcium/potassium ratio. This is often accompanied by a low chromium level. Hypothyroidism can be recognized by an increased zinc/copper ratio.

In the July 1977 issue of the *Journal of the International Academy of Preventive Medicine,* Charles J. Rudolph, Jr., D.O., Ph.D., illustrates and biochemically explains hair analysis patterns which indicate diabetes mellitus, hyperlipidema, hypertension, cardiovascular pathology, multiple sclerosis, schizophrenia, rheumatoid arthritis, and other conditions. We mention this (even though it is beyond the scope of this book) simply to indicate the extent to which hair analysis can be useful.

A Partial List of Laboratories Which Perform Hair Analysis

Analytical Labs Inc.
106 East Cheyenne Road
Colorado Springs, Colorado
  80906

Biochemical Concepts
6001 Marble N.E.
Albuquerque, New Mexico 87110

Da Vinci Labs
1 Executive Drive
South Burlington, Vermont,
  05401

Doctor's Data, Inc.
P.O. Box 111
West Chicago, Illinois 60185

Mineralab Inc.
22455 Maple Court
Hayward, California 94541

Parmae Laboratories
7101 Carpenter Freeway
Dallas, Texas 75247

If the hair analysis indicates the possibility of low blood sugar problems, we suggest that the tests for allergies be done before the glucose tolerance test. *Low blood sugar is frequently triggered by the ingestion or exposure to of an allergen.* When the person ceases eating (or touching or smelling) the allergen the low blood sugar attacks often disappear.

The Glucose Tolerance
Test and "Hypoglycemia"

The glucose tolerance test is used to assist in assessing a person's ability to maintain a normal blood sugar level. When a person's blood sugar level is frequently low, the condition sometimes is called "hypoglycemia." There is controversy in the medical community concerning hypoglycemia because some practitioners have used the name of the *condition* (hypoglycemia is a condition, not a specific disease) as a catch-all for numerous undiagnosed symptoms. In addition, there is a tendency for some neurotics to excuse their irrational behaviors by claiming that they have hypoglycemia (because the condition's symptoms include depression, irrational behavior, and fatigue). Perhaps all this has happened because hypoglycemia has no precise, simple cause. A chronic low blood sugar condition does exist, however, and it can cause a wide variety of negative, antisocial behaviors which, among other effects, can provoke discord in a marriage. A chronic low blood sugar level can also, in time, lead to other physical ailments. However, about 90 percent of the cases are relatively easy to control, and if they are controlled consistently an eventual cure is possible.

The causes of low blood sugar levels are many. Some of the most common are:

1. Allergies (there is growing evidence that allergies may be among the most frequent causes).
2. Stress.
3. Problems in the functioning of the pancreas.
4. Problems in the functioning of the adrenal glands.
5. Liver problems.
6. Inability of the digestive tract to absorb sugar at a normal rate.
7. Improper levels of hormones (such as glucagon) or improper levels of enzymes (such as sucrase).

(It is interesting to note that people who are allergic to tobacco smoke sometimes develop the symptoms characteristic of hypoglycemia, including outbursts of temper, physical weakness, irritability, and fatigue.)

Dr. Ross on the Glucose Tolerance Test

One of the nation's experts on hypoglycemia is Harvey M. Ross, M.D., a psychiatrist who specializes in treating depression. (Dr. Ross is president of the Academy of Orthomolecular Psychiatry and vice-president of the International College of Applied Nutrition.)

His book *Fighting Depression** includes a chapter on hypoglycemia and the glucose tolerance test, with details of the diet which Dr. Ross prescribes.

Dr. Ross gives full information (along with excellent graphs) on how to diagnose hypoglycemia from reading blood sugar levels during the five- or six-hour glucose-tolerance test. This information, of course, is available in the latest medical books on the subject. *But Dr. Ross also gives other diagnostic information which the client as well as the practitioner should know, and which is seldom available through standard medical books.* We quote from his book:

> Sometimes the glucose tolerance test is not sufficient to make the diagnosis. The patient's experience through the test must be

*Harvey M. Ross, M.D., *Fighting Depression* (New York: Larchmont Books, 1975).

weighed very heavily in determining the diagnosis. A glucose tolerance test that shows no abnormalities in the results of blood sugars is normal only *if the subject remains without any symptoms throughout the test.*

Sometimes I have seen test results that were normal, but the patient felt terrible throughout the testing and/or a few days following the test. Some of the symptoms that have been mentioned are tiredness, headache, irritability, hunger, depression, weakness, fainting, pallor, and sweating. When the history of having such experiences during the testing is obtained, there is enough evidence to presume the presence of hypoglycemia. When it is treated, the results are often gratifying.

This type of presumption is justified when one realizes that the testing is done at hourly intervals. I have seen tests done at half-hour intervals, where the blood sugar reached a low point at a half-hour and resumed the normal level at the regular hourly interval. In these tests, if only hourly intervals had been obtained, the level would have appeared normal and hypoglycemia would have been missed if the diagnosis was made on the basis of laboratory findings only. *The ideal way to conduct tests is with an experienced technician who is able to recognize symptoms and draw blood sugars at that time, rather than at a predetermined, specified time.*

The patient must keep a diary during the glucose tolerance test, noting her/his mood, physical feelings, mental feelings, concentration levels. When s/he feels low or uncomfortable s/he should immediately inform the laboratory technician, who will take an additional blood sample at that time.

Richard A. Kunin, M.D.,* has developed a glucose tolerance test form which the patient takes to the laboratory. It is an easy method for keeping the diary which is so important in making an accurate diagnosis. With Dr. Kunin's permission, the form is reproduced above. (It is one of a package of forms useful to practitioners which Dr. Kunin has created and copyrighted. These forms can be obtained from him at 2698 Pacific Avenue, San Francisco, California 94115.)

---

*Dr. Kunin is president of the Orthmolecular Medical Society and is author of the excellent book *Meganutrition* (New York: McGraw-Hill, 1980).

GLUCOSE TOLERANCE TEST 9

Dr. Kunin's Glucose Tolerance Test Form

NAME_____ DATE_____

Eat nothing after 9:00 PM the night before the test. Drink only WATER until the test is completed. No smoking. No coffee. No pills, not even vitamins, on the day of the test. Bring a book and stay at the laboratory during the test and read it in order to evaluate your concentration. Fill in the blanks as indicated below each time a blood specimen is taken: Do this on your own as the laboratory is not responsible for this. BRING THIS FORM TO YOUR NEXT APPOINTMENT.

	MOOD	PHYSICAL FEELINGS	MENTAL FEELINGS	CONCEN-TRATION	
	0	50	100	150	200
FASTING					
½ HOUR					
1 HOUR					
1½ HOURS					
2 HOURS					
2½ HOURS					
3 HOURS					
3½ HOURS					
4 HOURS					
5 HOURS					
NEXT DAY					

NOTE: This test requires an appointment; therefore call the lab to set a time.

RAKformGTT © Copyright 1978 Richard A. Kunin, M.D.

Before having a 5- or 6-hour test ordered, however [Dr. Ross continues], there should be a consultation with a physician to determine whether or not the symptoms for which the physician is being consulted are related to hypoglycemia.

Dr. S. Gyland studied several hundred patients with low blood sugar. He listed the frequency of symptoms as follows:

nervousness 94%
irritability 89%
exhaustion 87%
faintness, dizziness, tremor, cold sweats, hot flashes 86%
depression 77%
vertigo, dizziness 73%
drowsiness 72%
headaches 71%
digestive disturbances 69%
forgetfulness 67%
insomnia 62%
worrying and anxiety 62%
mental confusion 57%
internal trembling 57%
palpitation of the heart and rapid pulse 54%
muscle pains 53%
numbness 51%
indecisiveness 50%

Among the other symptoms which occurred in less than 50 percent of his patients were the following: unsocial, asocial, antisocial behavior, crying spells, lack of sex drive, allergies, incoordination, leg cramps, lack of concentration, blurred vision, twitching and jerking of muscles, itching and crawling sensations of skin, gasping for breath, smothering spells, staggering, sighing and yawning, impotence in males, unconsciousness, night terrors and nightmares, rheumatoid arthritis, phobias, fears, neurodermatitis, suicidal intent, nervous breakdown, convulsions.

In making a diagnosis, naturally not all of these symptoms must be present, but certainly some of the more common ones should be significant enough to be mentioned. Also, most of the time there is a history of a craving for or heavy intake of sugar, starch, or alcohol. If tremors or spells of weakness are prominent symptoms, these may be relieved by eating something sweet or starchy, and this is good presumptive evidence of hypoglycemia. All symptoms cannot be relieved by eating carbohydrates, as advised by an editorial in the *Journal of the American Medical Association.*

Because depression is so common in those with hypoglycemia, any person who is depressed without a clear-cut obvious cause for that depression should be suspected of having low blood sugar. As we noted about Dr. Gyland's studies, 77 percent of his patients with hypoglycemia complained of depression.

After the diagnosis of hypoglycemia is made, treatment can begin. Treatment is dietary, with the addition of food supplements. . . . I use large doses of B-complex, along with vitamins C and E.

We are not presenting the diet usually prescribed for the cure or control of the most common forms of hypoglycemia. The diet should be designed specifically for the individual after the medical practitioner has studied the results of the various diagnostic tests. *Every individual has unique dietary needs.*

APPENDICES

APPENDIX I

"No Man Is an Island": The General Systems Theory

According to the systems concept, the whole is *more* than the sum of its parts.

The whole consists of all the parts *plus* the way the parts operate in relation to one another. For example, a human relationship (which is a system) consists of a continuum of behavior exchanges between the people involved. In a childless marriage, there are the wife's behaviors and the husband's behaviors. However, once the behaviors are exchanged, a "system" results. This system (the relationship) is the pattern of reactions, one to the other, which becomes a relationship that is more than the simple sum of its parts (the individual behaviors).

The systems concept postulates a constant action-reaction among all living things. Some of the action reactions are so weak that they are imperceptible. The closer the association between things, the more obvious is the action-reaction. Once an action-reaction pattern is established, it tends to endure. It resists change. *If an influence upsets the established action-reaction*

*Adapted from William J. Lederer and Don D. Jackson, M.D., *The Mirages of Marriage* (New York: W. W. Norton, 1968), pp. 87–97.

balance between the associated entities, then a compensating factor is provided by the system, to restore balance. For example:

A person is a system of bones, muscles, nerves, etc. If a person goes blind, the system is thrown out of balance. Unless medicine restores the sight, the system must provide a compensating factor. What is the compensating factor? A person who goes blind soon acquires extraordinary facility in hearing. A blind person uses sounds almost like radar, can teach her/his fingers to be sensitive enough to read Braille, and develops a more acute sense of smell.

Another example: The kind of interaction referred to by the systems concept is illustrated by the body's mechanism for maintaining the right amount of fluid and the proper body temperature. Normally, a certain amount of perspiration comes to the skin surface, where it evaporates and thus cools the body. The warmer the body, the more it perspires, in an effort to maintain a temperature of 98.6°.

As perspiration leaves the body and evaporates from the skin, fluid is lost and must be replaced. But perspiration also contains salt, and salt, too, must be replaced, not only because the body needs a certain salt level but because water cannot be retained unless salt is present. In turn, salt cannot be retained unless small amounts of potassium are present. Under these circumstances, the individual normally craves foods which contain both the salt and the potassium required, thereby providing the necessary compensation to keep the system in balance. Salt balance, potassium balance, fluid balance, body temperature, perspiration process, desire for salt, and thirst are all directly related to each other. They are part of the system.

The system is the sum of all parts plus the way the parts interact with each other. Once a system is established, it resists change. If *one part* of the system changes, the system can be kept in a steady state either by (1) *all other parts changing to keep the system in balance,* or (2) the system bringing the part which initiated change back in line.

For example, a barbershop quartet is singing in perfect harmony. If the first tenor changes the key or the beat, the other three frantically try to get the tenor to come back to the original key or beat. If they are not successful, the system is out of balance. Either the other members must get the tenor to change back *or they must all change their key or beat so that the system is back in harmony (balance) again.*

A "system," in human life, is a group of interacting people. The behavioral pattern of the group (the whole) is different from the various behavioral patterns of the individual group members. The behavioral pattern of the group is determined by the way the individuals in the group relate to one other, plus the way each relates to the group. *A change in behavior of one person in the group affects the behaviors of every other person in the group as well as of the group itself.*

In a marriage, the systems concept operates as unequivocally as it does in the examples given above. To grasp this, one must accept the fact that in its totality, *marriage is not a rigid and static relationship between two rigid and static individuals.* Marriage is a fluid relationship between two spouses and their two individual systems of behavior. The totality of marriage is determined by how the spouses behave *in relation to each other.* Put John Smith alone on an isolated island and he will have one behavioral pattern. Put John in an office with a harsh boss and he will have another. Put John in the intimate relationship of marriage and he will have still another, one largely depending on his response to the behavioral pattern of his wife, just as she is constantly responding to his pattern.*

It is obvious that human behavior is seldom constant for long. An individual may be cheerful one morning, irritable the next. He may be generous and benevolent one afternoon, stingy and mean the next. *The systems concept makes it clear that a change in the behavior of one spouse is often a reaction to changes in his partner's behavior; the reaction, in turn, causes additional changes in the partner's behavior.*

This action-reaction system operates in a *circular* fashion (sometimes constructive, sometimes destructive). However, changes in a system can also be introduced by factors *outside* the system, such as the death of one spouse's parent. In such

---

*For simplicity, we are here discussing a childless couple. When there are children, they are just as powerful in the relationship as the parents.

cases, the unilateral change in one partner must be dealt with and incorporated into the family system. The other spouse will respond (perhaps without being aware) in a manner designed to keep both her/himself and the system in balance.

The marriage system appears even more complicated when one realizes that whenever a person is in the company of *any* other person, a new system is generated. When husband and wife go out together and meet other people, their behavior patterns are different from when the two are alone. The presence of others feeds new stimuli into the joint wife-husband system, which becomes a *new system* as long as the other persons are present. Husbands and wives often wonder why they behave differently when they are in public than when they are home alone, or why they behave differently when they are together with their children than when each is alone with the children. This apparent inconsistency often causes trouble. Each spouse notices the changes in the other but not in himself. An objective observer would see that they both change. These alterations are unavoidable because all additions to or departures from a group in some way change the system of the group as a whole. (A spouse alone in a therapist's office behaves differently than when in the office with the other spouse. They both behave differently if their children are also in the office, and all of them behave differently in the doctor's office than at home.)

The systems concept applies to everything, from the atoms of inanimate objects to the relationships among people. Once a person is conscious of this, s/he has the ability to ask her/himself: "If I behave in such-and-such a manner, what will happen to the system? How will the tone or the behavioral pattern of my marriage be altered?"

In a marriage, each partner tries to maintain behavioral systems which provide him/herself with maximum satisfaction. Sometimes the satisfaction assumes neurotic dimensions, such as finding pleasure in illness or in incompetence because they can be used as a weapon against the spouse. When *both* partners are in a state of satisfaction, there is a positive emotional and psychic balance. That is what they strive for. However, human behavior changes frequently, even radically, and every action and mood of one spouse begets a reaction from the other.

Therefore, to remain in balance, the marriage system is always in a state of flux. The forces in it move this way and then that way, go up and down to various levels, increase and decrease in intensity. The systems concept makes this situation clear.

When two people marry, the first important action is the attempt of each spouse to determine the nature of the relationship: each spouse wants the system to be satisfying to her/himself and would prefer to achieve this end without changing her/his established behavioral pattern. Each wants the other partner to make the accommodations. Usually a spouse approves of his/her own ways of behaving, her/his own mannerisms, habits, and performances, and finds fault with those of the other. For this reason almost all marriages—at least at first—contain friction. To reduce this friction is often difficult because of "behavioral blindness": neither spouse is aware of how others perceive her/him. Nevertheless, each attempts to shape the relationship, to influence how the joint system will operate, and to determine the limits of acceptable behavior.

Once a system is established, it tends to remain static. When a man and a woman are first married, they exchange a wide variety of behaviors. After a time—from several months to several years—they try to work out mutually acceptable ways of labeling and interrelating their behavior so that each individual feels s/he is an equal. This may result in some incompatible behaviors. In a workable marriage, the maladaptive, incompatible behaviors die out in ways which do not upset the system.

The conclusion, of course, is that *when all components of a system (in this case marriage) participate equally in the effort to effect change, then the system does not resist the change.*

In this book there has not been much mention of the system, even though the principles have been included in the exercises. The object, of course, is to change the present system to one which provides greater satisfaction to both partners in the marriage. Readers must, therefore, begin by working on the system as it now is. The current state of the system is termed the "norm." The norm consists of the marriage system as it has been functioning up to the present moment. The current norm is the old pattern of interactions which the spouses have accepted and expected, *whether they are desirable and pleasant or discor-*

dant and unpleasant. It is the book's aim that the spouses learn how to change that system, to create a new and more desirable norm, after first agreeing on a definition of the *new* norm will be.

To keep the marriage in balance, the spouses must learn to identify the inevitable changes which will, over time, occur in themselves, their marriage, and the circumstances around them. After identifying these changes they must learn how to adapt to them. Indeed, they must *constantly* adapt to changes in their lives and marriage by constantly (on a monthly or yearly basis) agreeing to new behavioral contracts.

APPENDIX II

Notes on Sources,
Background Information,
Abstract of Clinical Procedures,
Professional Comments

Information in this book was largely obtained from the following sources:

1. Published studies, together with the author's clinical experiences with (a) married couples who enjoyed stable and satisfying relationships; (b) married couples who sought help in improving their troubled relationships; (c) couples seeking premarital advice; (d) divorced people, female and male, who wanted advice that would help them in avoiding repetition of previous relationships.

2. A research project conducted by Dr. Don D. Jackson and W. J. Lederer (1961–65). The samplings consisted of 278 couples about to be married. Approximately half of them had been living together for about a year prior to the wedding. The couples were interviewed in depth before their weddings and after their honeymoons. For the next four and a half years they were questioned on a regular basis. This research was an important

source of the data and interpretation to be found in Jackson and Lederer's book *The Mirages of Marriage* (New York: W. W. Norton, 1968).

3. Work done at the Behavior Research Institute (Cambridge, Massachusetts, and Peacham, Vermont)* with couples having marital and family problems. Various therapies were used, depending on the case, including video-therapy, transgenerational differentiation, behavior modification, modified analysis, orthomolecular (nutritional) therapy, and clinical ecology, frequently in combination. Because of the Institute's belief in the systems concept much of the therapy was carried out in clients' homes with all family members present, and usually with the use of videotape.

An abstract of clinical procedures follows.

An Abstract of the Clinical Procedures Used at the Behavior Research Institute

1. The first interview with clients was held in our office. At that time they wrote out their family histories for three genera-

---

*The director, W. J. Lederer, studied preventive medicine for rural areas at the University of Szechuan, in China, where he received, in addition to a Ph.D. in Oriental studies, an M.D. degree in rural preventive medicine; he is not a physician with a license to practice in the United States. His full-time work is research. His activities have been directed toward (1) nutrition, (2) environmental forces which influence behavior, and (3) the prediction of social group responses to political, economic, and social influences. In the 1950s his research was directed at the problem of why the strong, affluent American nation was suffering dramatic political and economic defeats all over the world, especially in Asia. This research produced the books *The Ugly American, A Nation of Sheep, Our Own Worst Enemy,* and *Sarkhan* (reissued as *The Deceptive American*).

One of the symptoms of a nation's decline is the breakdown of the family and marriage. The frequency of these breakdowns in the United States prompted Lederer (with Don D. Jackson, M.D.) to research the structure and general nature of marriage. This produced *The Mirages of Marriage.* Since then Lederer has been formulating behavior modification procedures by which couples can repair and improve their long-range intimate relationships. This work has resulted in much of the present work, *Marital Choices.*

At this writing, Lederer is beginning to investigate the means whereby an individual can be fully functional and joyful even when living in a discordant environment such as present-day society.

tions, including divorces, sicknesses, death dates, etc. They were give a Relationship Assessment Test and the HOD Test for schizophrenia.* If the HOD Test indicated schizophrenia, the client was referred to an appropriate therapist.

2. It was arranged for clients to have physical diagnostic tests, including a normal check-up, a full blood profile, a hair analysis, a urinalysis, an allergy test (which most clients did at home, using the Coca Pulse Test), and a five-hour glucose-tolerance test. Those clients who preferred to use a clinical ecologist or allergist were referred to one.

3. When needed, a nutrient program was prescribed to bring clients into biochemical balance.

4. It was explained to clients that all psychological therapy sessions would be held in the client's home and that it was mandatory for every member of the household to be present at every session. If one person were absent, the session would be called off.

5. We went into clients' homes with portable video equipment. There was at least one session "for fun." This was a social event, not a therapy period. Everyone in the family was taught to use the video equipment, and everyone made "movies" of each other. These were played back for them to enjoy. This procedure was used to acquaint the entire family with the equipment so that each of them, including the children, knew how to operate the video camera. Also, it served to make the practitioner a part of the family system and permitted her/him to observe how the family members interacted under conditions of minimum stress. The "fun sessions" usually lasted four or five hours.

6. About a week later, the practitioner, with video equipment, went into the home for the first therapy session. The visit began about an hour before the evening meal, to permit the therapist to help out around the house and reinforce his/her status as a member of the family system. (We encouraged a first-name relationship.) The actual work began during the eve-

---

*The Hoffer-Osmond Diagnostic Test (for schizophrenia), available through Bell Therapeutic Supplies, 196 Rockaway Avenue, Valley Stream, N.Y. 11581; (516) 561-7665.

ning meal, with all members of the household present. (Note: The evening meal is mankind's oldest ritual. There is the least "acting" or lying at that time. Unless there are strangers present, the family usually behaves as a "closed system.")

When the meal started, the practitioner began videotaping the family. After three or four minutes, the therapist said that s/he was hungry and asked a family member to take over. The rest of the videotaping was done by the family members, with everyone, including the small children (who sometimes needed assistance in operating the camera) having her/his chance. *They invariably would film the behaviors which they did not like in others.* By the end of the meal the therapist had a documentary film of the family's undesirable behaviors as perceived by the individual members. This was played back after dinner. Family members were asked whether what they had just viewed represented the way they wanted to live. Everyone, including the small children, was required to answer this question.

The next question concerned *how* they would like to live. In short, the family members were asked to define their objectives, to define the behavioral changes they wanted. This discussion also was filmed and played back.

Then the family members were asked to discuss a schedule of self-prescribed behavioral changes. They were informed that they had six weeks in which to accomplish these changes. If they did not succeed, they would have to go elsewhere for help.

It is important that the therapy be time-bound.

This type of video session was repeated once a week for the next six weeks. At each session the previous week's tape was played back, as well as the current week's tape. Family members were requested to identify any progress which had been made and to define the progress they wanted to make in the following week. These sessions usually lasted four to six hours. At the end of each session the family was given assignments to work on during the next seven days, similar to the ones in Part I of this book.

7. The program (both the nutritional and the behavior modification elements) requires monitoring, either by family members or by the therapist, if a therapist is used. We required one

member of the family to telephone us every morning at a specific time to inform us of what good things had happened in the family during the previous day and what the caller had done to help in making them happen. All family members, including the young children, took turns making the calls.

Modifying behaviors under the above circumstances is relatively simple when dealing with normogenic families (we did not work with schizophrenic or paranoidal families). Families can manage behavior modification alone if the psychological exercises such as those provided in this manual are available to them. The more difficult task is developing and maintaining the part of the program concerning biochemical balance. Those who were in biochemical imbalance (57 percent of the total number of clients) usually made good progress at the beginning of the program. However, those who violated their nutritional regime or their allergy-elimination process after a few weeks, fell back into their discordant behaviors more often than did the others.

For those who adhered to the entire program, the improvement in the relationships appeared to be permanent in an unusually high percentage of the cases. This conclusion is based on a follow-up done six months later. We have no information on what happened six months after that

Some people who use this book will opt to work with professional therapists. For them we have attempted to create a therapy session which does not require a therapist to put in the long hours we found necessary for our research, to eliminate the need for expensive video equipment, but still provide clients with perceptual experiences similar to those made possible with video equipment. To eliminate the need for video equipment we created the exercises found in the body language assignments. In some ways these assignments are considerably superior to the video routine, because families can achieve the same results on their own, each member helping the others to perceive themselves as others perceive them.

The book is designed so that the practitioner who wants to use *Marital Choices* can assist a great many clients at the same time. The practitioner must conduct the initial interview, arrange the diagnostic tests, prescribe a biochemical balancing program, and then assign the clients to the approximately six weeks of exercises described in this manual. If it is desired to have the clients' progress monitored, the monitoring can be done in the home by nurses, social workers, members of the clergy, or friends of the clients.

When the book is used by therapists, its methods are advantageous both to the clinicians and to their patients. The therapy has a high success factor, is relatively brief, and is much less expensive than traditional methods.

The possibility of increasing human functionality through proper nutrition and supplements of minerals and vitamins are suggested by a recent research paper by Harrell, Capp, Davis, Peerless, and Ravitz (*Proceedings, National Academy of Science, USA,* vol. 78, no. 1 (January 1981), pp. 574–578).

The research was done on mentally retarded children. However, we at the Behavior Research Institute have observed a similar improvement process in both adults and children of discordant families composed of normogenic individuals. It is for this reason that *Marital Choices* recommends diagnostic tests which will lead to prescriptions for adequate nutrition and supplements of minerals and vitamins. In addition to this we prescribe modifying the behavior (from negative to positive) of all parties within the relationship. We believe that in the joyful and healthy relationship body, mind, and spirit go hand in hand. That is what we have attempted; and that is why the manual is called "holistic."